ETHEREUM FOR

BEGINNERS

A Simple Complete Guide to Investing in the New
Cryptocurrency Ethereum

(Complete Guide to Ethereum and the Blockchain
Technology)

Robert Fisher

Published by Tomas Edwards

Robert Fisher

All Rights Reserved

Ethereum for Beginners: A Simple Complete Guide to Investing in the New Cryptocurrency Ethereum (Complete Guide to Ethereum and the Blockchain Technology)

ISBN 978-1-990373-68-8

Legal & Disclaimer

The information contained in this book is not designed to replace or take the place of any form of medicine or professional medical advice. The information in this book has been provided for educational and entertainment purposes only.

The information contained in this book has been compiled from sources deemed reliable, and it is accurate to the best of the Author's knowledge; however, the Author cannot guarantee its accuracy and validity and cannot be held liable for any errors or omissions. Changes are periodically made to this book. You must consult your doctor or get professional medical advice before using any of the suggested remedies, techniques, or information in this book.

Table of Contents

Introduction

Cryptocurrencies have captured the attention of investors and traders all over the world. One of the most successful digital currencies currently being traded is ethereum. It is an online currency that has seen its value grow exponentially in the last one year. It is based on a solid, secure and incorruptible blockchain system that provides a foundation for running smart contracts on it.

Ethereum is quite different from bitcoin, which is the digital currency system that most people are used to. It is not a currency as such but a system that uses ether as a currency. There has been a lot of attention on ethereum lately, prompting interested investors to buy this digital cash in the hope that it will continue its phenomenal growth in the years to come.

This book explains in great detail all about ethereum, how it works, and the best ways of investing in it and making money. This book also looks at cryptocurrencies in general and why they are bound to affect the way transactions are handled in the economic and finance world. Digital currencies and the blockchain have taken the world by storm and are changing the way things are done. Ordinary people are benefiting from these currencies and the promise that they hold. You too can learn and profit from cryptocurrencies.

Chapter 1: What Are Cryptocurrency, Bitcoin, Ethereum And Money?

Cryptocurrency! What is it? Using the encryption techniques of cryptography and cutting edge computing it is possible to create money that is called cryptocurrency. Money is a store of value; it must be transferable and transfers must be capable of verification.

Cryptocurrency has these features. It also has something else, which has significantly enhanced its appeal, for many, and that is its complete independence from any government or central bank. Cryptocurrencies are decentralized, with no central authority controlling them.

Recently, most large banks have become involved with the technology that has allowed cryptocurrencies to flourish. However such currency as comes from them will be controlled by these banks and will not be cryptocurrency. The original creators of cryptocurrency were insistent there must never be such centralization.

The development of cryptocurrencies:Bitcoin was the first cryptocurrency and has been the subject of much publicity; its value in dollar terms has considerably increased in the last couple of years. As well as Bitcoin, there are a lot of other cryptocurrencies, one of which, Ethereum, is the subject of this book.

Many of these cryptocurrencies have also had a great increase in their dollar value.Government and legitimate business have an ever-increasing involvement in cryptocurrency and there is a huge market for these currencies, with the market capitalization of all cryptocurrencies in late August 2017 more than $150,000,000,000! Facts about cryptocurrency: During the latter part of 2017, there were more than 1100 cryptocurrencies. Bitcoin, the cryptocurrency with greatest market capitalization, has received much unfortunate attention because recent hacking attacks on organizations around the world have demanded Bitcoins as ransom. Many in the criminal world believe payments in cryptocurrency are untraceable. In fact, this is only partially true.

If you want to see all sorts of intriguing things about cryptocurrency, then log onto the website coinmarketcap. Each of the top 100 cryptocurrencies, by market capitalization, has a little graph next to it, revealing the fluctuation of its value, in US$, during the last seven days. You can also find the change in the value as a percentage in the last 24 hours. If your interest concerns all cryptocurrencies, then there is another table showing a great many more cryptocurrencies, with the same information, but minus the graphs.

Another site, with valuable information about cryptocurrencies, is worldcoinindex. This site is similar to coinmarketcap. However, it is not identical and does not have the same information but should be a reference for anyone interested in cryptocurrency.

Checking one of these shows that Bitcoin (BTC) has a large proportion, nearly 50% of the total capitalization. Its actual market capitalization was more than $70 billion (US). One BTC had a value during August 2017 of $4300. Other cryptocurrencies also had large market capitalizations and values.

Ethereum (ETH), the subject of this book, had market capitalization greater than $30 billion and one Ether (ETH), the coin of Ethereum, had a value more than $330. Not all cryptocurrencies by a large market capitalization have coins of high value; an example is Ripple (XRP) whose rank in market capitalization is about fourth with a value of $8 billion. However, 1 XRP only had a value of about $0.2 (US) at that time.

The reader may think that all cryptocurrencies have high market capitalization. This belief is not true, as a check of all cryptocurrencies shows. At the same time that Bitcoin, Ethereum, and Ripple had the values listed, another cryptocurrency Digital Coin (DMB) had a market capitalization $455 (US), and there were many others much lower than this. We will discuss this myriad of coins and how they arise later.

How is cryptocurrency created? In the first paragraph, we said this creation used the encryption techniques of cryptography and cutting edge computing technology. Expanding on this, we can compare a cryptocurrency to an incorruptible digital ledger. All financial dealings in this currency are recorded in the ledger.A large number of powerful computers called nodes maintain this ledger. The network of nodes maintains the blockchain and the next chapter has much greater detail about it. We call the process of producing cryptocurrencies mining.

Mining of cryptocurrencies: It is an intriguing fact that there will never be more than 21,000,000 Bitcoins. When Bitcoins are made the process by which this occurs is called mining. The initial Bitcoiners, Satoshi Nakamoto and Hal Finney, were able to mine Bitcoins using desktop computers, but as more and more

Bitcoins were produced mining required computers of much greater power.

This problem is caused by the use of hash functions, which are mathematical objects that are very difficult for lay people to wrap their heads around and discussed far more thoroughly later. Most cryptocurrencies arise from similar mining and this includes Ether, the coin of Ethereum, subject of this book.

Unlike Bitcoin, the number of Ethereum is unlimited, although the total number of Ether possible each year is about 18 million. However, Ethereum does not have a hard cap, which is a maximum number of coins. This lack of a hard cap may cause problems in the future.

Cryptocurrency, stocks and ordinary money: With the word cryptocurrency used to signify electronic money, the reader may wonder what the term is for the currencies in our daily lives such as the dollar, euro (Europe) or renminbi (China). The term used by those involved with cryptocurrency is fiat currency.

In some ways, cryptocurrency resembles the stocks and shares of the stock market than money, as most of us are accustomed to using. Currently, most cryptocurrencies need purchase from cryptocurrency exchanges in the same way that someone buys stocks and shares from a stock exchange. The acquisition of a cryptocurrency is the buying of part of the digital computer network for that currency and an entry in the blockchain of that cryptocurrency.

Is cryptocurrency different to money? Earlier we said that money is a store of value that it must be transferable and the transfers must be capable of verification. Cryptocurrency has these latter two features. However, if you have looked at coinmarketcap, you would see great fluctuations in the value of all cryptocurrencies. It is probably fair to say that cryptocurrency is not yet a reliable store of value.

In countries, which are politically or socially unstable, the currencies are not reliable stores of value. Most cryptocurrencies resemble the currency of unstable countries with the worth not consistent. In 2017, most cryptocurrency is bought for speculative reasons and with the high increases in the values of the major cryptocurrencies, this is not a serious problem. However, a significant downturn in the values of these currencies could cause a crisis, if too many people invest too great a proportion of their wealth in cryptocurrency.

Does cryptocurrency have advantages over fiat money? These have been present from Bitcoin's inception.

• Cryptocurrencies, usually have the benefit of complete decentralization. No nation, business or bank control them.

• Cryptocurrencies rely on their blockchain and its nodes. Failure of some nodes will not prevent the continuation of the currency. This reliability is similar to the Internet.

• When most cryptocurrencies are used in transactions, there is greater anonymity and privacy than using fiat currencies. However, having said that, there is now a belief that cryptocurrencies are about as anonymous and private as cash. Governments are using some of the best computing experts to find ways of unraveling transactions in cryptocurrency.

• Initially, cryptocurrencies may seem strange. However, usually, with practice in using them, they become as simple as fiat money.

• Cryptocurrencies are, for many, much cheaper and quicker than fiat currencies with their fees and delays in the completion of transactions.

• With cryptocurrency, there is no danger of charge back, where a dissatisfied customer is allowed cancel a transaction that involves credit cards. This lack of chargeback is a feature of great appeal to traders.

• A useful feature of cryptocurrencies is that as a result of their electronic nature they are far more robust and portable than cash and notes.

• Fraud is much less probable with cryptocurrency than with fiat money, which, sadly, now suffers from frequent theft by hackers of credit card numbers and subsequent sale on the Internet.

Blockchain: We discussed this previously and just mention that when we write blockchain, we are using a shortened version of writing blockchain technology. By use of this, the distribution of digital information, without the threat of being copied, is possible. When it first began blockchain was only for the use of Bitcoin, which was the first and is still the largest cryptocurrency.

Blockchain, as with all new technologies, has new uses being discovered for it every day. Initially blockchain, through Bitcoin, was visualized as a way to overcome the deficiencies of the global banking system that nearly caused a financial meltdown in 2007-2008.

Nowadays, we know that this technology has applications far beyond money and finance. Ethereum fulfills this realization, and future chapters of this book will illustrate how this amazing technology is doing this.

Chapter 2: Ethereum Coin - What Is It?

You might now be asking yourself about Ethereum coin, and what it is, and that's a very valid question. What makes it so different from all the other cryptocurrencies and altcoins that are out there on the market?

After all, with Bitcoin leading the pack as the first coin on the market, and a slew of other coins that you could invest in, what makes Ethereum special enough to stand out of the pack and be considered something special?

The answer to that is pretty simple. What makes Ethereum better is how it's utilized on a grand scale, and how it enables users to intuitively have control over their pool of Ethereum wealth that no other cryptocurrency can compete with.

History

To put it in perspective, let's take a little bit of a look back at its history and where it got its start.

Back in 2013 a programmed named Vitalik Buterin who was involved with the Bitcoin project when it first got off of the ground believed in a decentralized coin that could actually be continued to be expanded on. At the time, part of the problem Mr. Buterin was seeing with bitcoin was it was becoming mainstream, and accepted slowly as an alternative to conventional methods of holding wealth. While initially, this was the goal with Bitcoin, the fact that it could become completely centralized under a government control was part of the problem, he felt and severely limiting what Bitcoin was supposed to initially revolutionize.

So in March of 2014, Ethereum was being fully developed and was brought on board as a project that could have the potential to extend blockchain use (The main way that "miners" mine for bitcoin or other cryptocurrencies) beyond the traditional peer-to-peer limitation of the altcoin coding. While legal issues and questions arose to both its legality what Ethereum could possibly even do, it soon became apparent that it was a legitimate project after Buterin won the "World Technology Award" for the creation of Ethereum.

Soon after, crowdfunding for the project began to start pouring in in July to August of 2014, with initial investors pooling in their bitcoins to purchase in Ethereum coins. The end of the crowdfunding event happened on September 2nd, 2014 at the conclusion of nearly $20million in sales generated from just Bitcoin purchases alone,

By the end of May 2016, Ethereum market value quickly rose to more than 1 billion USD. It was quickly becoming a serious contender against Bitcoin itself, mostly due in part to the various different services that the Ethereum platform promises that Bitcoin itself cannot do. Today, it's still being developed on and maintained, but each new service that's created centralized around Ethereum continues to prove it's staying power over other competing Altcoins who can't keep up.

Applications

So now you have a bit of understanding of where it came from, what can it do? What applications are being talked about that make it worth investing time and money into?

As well as being a traditional cryptocurrency that you can hold on you like any other one on the market, the Ethereum platform has a wide range of different uses to them other than simply being a digital object that you can possess. In hypothetical uses, Ethereum is mostly used with higher end software that can be used to establish online marketing platforms, where instead of traditional uses where you open a store, tie it to your bank and tax information and social security and stuff, you simply open up a trading wallet, and trade Ethereum for goods and services quickly and efficiently.

In other uses, it's seen mostly in having a way to enjoy having access to a decentralized monetary system that's not tied to any government oversight. This allows Ethereum to be traded for a wide range of different objects, even physical ones. It's not unusual to see it quickly used for projects in finances in lieu of physical currency, identity management, electricity and resource allocation, gambling betting, and even used for arts, crafts, and even food. In fact, some musical artists like Imogen Heap and other independent artists used Ethereum trading for their music in lieu of traditional currencies.

Because it's easy to work with, easy to use, and easy to trade you can even use it to bypass business models that normally would be too costly to run as well. With Ethereum, you don't have to worry about import/export taxes, sales taxes, and you can even create smart applications and contracts that won't require a lawyer's fee as well to understand.

Smart Investment

So with all of those choices, applications, and information that was previously shown, how widespread is it? After all, it doesn't really do any good if anyone invests in it, but you still can't use it for anything other than the group of people who still have it. It's no more valuable than owning a handful of Carnival Tokens. They look pretty, but they have limited application.

True, while the New York Times noted in march of 2016 that the Ethereum platform adoption is still in its early phases and mentioned that its complexity might be too ambitious, the fact that various companies, institutions, and other organizations having faith in the Ethereum platform gives more credibility to the cryptocurrency over its other competitors, even over Bitcoin.

in fact, Microsoft announced a partnership with ConsensSys, a blockchain startup focused on the Ethereum technology, to help develop a cloud-based computer business model to allow people who use cryptocurrencies to trade their coins for Microsoft services, including securities, OS applications, cross-border payments, trading, and more. Even allowing a way for the technology to be used within the Microsoft Visual Studio programming packages as well.

Not only that, but ConsenSys is also using the Ethereum platform as well to help create by 2017 a fully functioning digital bank to be allowed to store, trade, and even invest in Ethereum and other cryptocurrencies as well.

While there are still a lot of initial problems to the platform itself, it's successes have outweighed heavily the negatives, and shows that investing in Ethereum is something that's considered a smart choice for the future.

Chapter 3: What Are The Applications Of Blockchain Technology?

You do not have to know how exactly Blockchain technology works, just realize that it is going to be very important in the future. At the current time, there is well in excess of $500 billion US in money transferred each year, which currently involves billions of dollars for middlemen and millions of wasted hours. Blockchain technology will completely eliminate this.

Enhanced security through the Blockchain via Cryptographic Keys

A lot of the wasted time is currently spent in identity verification. With Blockchain technology there will be a decentralization of online identification. The risk posed by data being held in a central location will be eliminated by the Blockchain storage of data throughout its network. There are no centralized, vulnerable points that computer hackers are able to exploit.

The foundation for this is what is called **public** and **private keys**. A public key is a very large number that, on the Blockchain, is the address of the user. When a Bitcoin or other cryptocurrency is transmitted across the network then it gets recorded as being at that address. The **private key** is a password giving the owner access to their cryptocurrency or other assets in digital form. By storing data on the Blockchain it is rendered incorruptible. The only vulnerability will store the private key in what is called a paper wallet. Currently, this involves the printing of the key in a form called a QR code, familiar to users of apps.

A new network? Web 3.0?

There is a new layer of functionality on the web as a result of Blockchain technology. The widespread use of Bitcoin has demonstrated the power of this model. The great security afforded by this technology will cause the complete transformation of the business of finance. Savings as a result of this transformation in clearance and settlements will be in the billions of dollars.

This technology will facilitate the following.

Smart Contracts

We will have much more to say about these later. Think of them as distributed ledgers that allow the simple contracts to be completed once certain conditions are reached.

A Peer-To-Peer Economy

This means that this technology will enable two people to do business directly rather than through an intermediary. There will be apps that will harness this technology to bring this into being. For example, OpenBazaar uses Blockchain technology to produce a market like eBay that is peer-to-peer.

Crowdfunding

There are moves to use this technology to fund the development of new products. An example of how this might work was the raising of $200,000,000 for the Decentralized Autonomous Organization (DAO). Unfortunately, there were major problems with hackers however the basic idea was shown to be viable.

Stock market

Using Blockchain technology share trading could be revolutionized with settlement in seconds instead of the current three days. Trading would become peer-to-peer and the whole process would be transformed. The application of this technology to stock markets and exchanges is being trialed throughout the world including Australia, Japan, Germany and the USA.

Polls and Elections

Some see this technology as making elections and polls far more transparent than they currently are. There is already an app called **Boardroom** which uses Ethereum used in this area.

Checking Veracity of Claims

Increasingly customers want to know their goods are produced ethically and not with child labor, misuse of resources, genetic modification, slaughter of endangered species etc. It is quite easy now for rogue suppliers to provide false information to assuage the consciences of concerned customers. By use of Blockchain technology, the supply chain would be largely free of this sort of corruption. Ethereum is already being used in a number of situations and is certain to be involved in much more.

File Storage

With its distributed network, the task of safely storing data is greatly improved. It will also make the transfer of data much more efficient.

Electricity and the Internet of Things (IoT)

The involvement of Blockchain technology in the distribution of electricity seems almost unbelievable however with the move to solar energy throughout the world Ethereum based smart contracts are redistributing excess solar energy and similar contracts are envisaged as the application of Internet technology is spread to many other important human activities in what is called the **Internet Of Things (IoT)**.

Intellectual Property Protection

The use of Blockchain technology will make the protection of intellectual property such as films, music, and books far stronger than is currently the case where it is quite simple for the determined to obtain pirated copies of works with copyright.

Identity

Currently, most people who use the Internet a lot have or should have a notebook with all their logins and passwords. As hackers have become ever more able passwords have had to become longer and more complicated.

Few people realize the huge complications arising from eCommerce and what goes on behind the scene. Blockchain technology offers a means by which this process could be made far more efficient.

Disrupt Money Laundering

One of the first things criminals have to do with their ill gotten gains is disguise where the money arising from the sale of weapons, drugs, endangered species etc. comes from. Currently, they are able to do this via gambling, shell companies, false bank accounts and other forms of subterfuge. The use of Blockchain technology will make all these almost impossible.

Management of Personal Data

At the moment organizations such as Facebook, Twitter, and other social media have free access to the personal data of their users and make many billions of dollars with it. By use of Blockchain technology, users will be able to sell their data. There are trials of such ideas emanating from the great American university MIT. A new economy dawns!

Registration of Land Titles

Throughout the world, these are a problem for everyone from owners to developers. Dealing with these is time-consuming and expensive; prone to fraud and corruption. Fortunately, Blockchain technology offers an escape from this turgid morass. Throughout the world trials of this technology are underway to realize this.

Prediction Markets

This is a concept that is not part of most people's everyday experience. Such markets depend on mathematical probabilities combined with market demand. While still in its infancy this idea is being developed and uses Blockchain technology.

Chapter 4: The Ethereum Virtual Machine

The Ethereum virtual machine can be defined as a world computer that can execute code from any computer in the world. It is, in essence, the heart of Ethereum. The Ethereum virtual machine allows the safe running of code that varies in complexity and language used. The computer executes untrusted code and provides protection to the applications created by programmers all over the world.

One of the most important uses of EVM is the running of smart contracts. These types of contracts are becoming more and more popular, especially in the world of crypto-currencies. One thing that they do is eliminate is the problem of denial-of-service attacks. These attacks are when the intended users cannot access services due to a hack on the host's network.

The virtual machine is, in one way or another, a protector of the applications.

The language used by developers is straightforward, unlike the language that is needed to expand an existing platform or build one from scratch. The use of existing programming languages, such as JavaScript, means that even the average mobile application (or 'app') developer can do the same on Ethereum. It is much easier to create an application on the Ethereum Virtual Machine in comparison to the other open-source blockchain platforms.

The virtual machine is a way for the programmers to reach out to each other in cyberspace and share the application that they have created, no matter where they are in the world. The only necessity to access the EVM is to have a computer and access to the internet. The machine is said to be Turing complete. When a device is said to be Turing complete, it means that it can complete any algorithm that has been presented to it, despite the complexity. Before the invention of Ethereum, a Turing complete computer was more of a theoretical term. The use of blockchain technology and the diversity of the Ethereum Virtual Machine has seen this become a reality.

It would be costly and very slow to run a world computer if it was in a central position. The logistics of the whole process of building it and running it would be very taxing and expensive. The Ethereum Virtual computer has made all this easier by using the contributions of all the people on the network in order to solve the problems that would face a machine in a central position.

The Ethereum Virtual Machine (EVM) can be understood as the perfect testing ground for decentralized applications. The EVM is sandboxed from the rest of the network. This makes it a haven for programmers that can create the applications which are based on Ethereum.

The runtime environment is ideal for those who want to use smart contracts. Various businesses can run using smart contracts, and the Ethereum machine makes this possible. Applications can be created that are independent of any government, legal or social system. This ability makes the exchange of items or currency much faster and cheaper. The transactions are much cheaper compared to the transaction fees charged currently, especially on international exchanges.

Unlike the previous applications built on blockchain technology, such as Bitcoin and other crypto-currencies, you can easily share the programs you create on the Ethereum network instead of just sharing the data. This way, instead of just sharing data, you can also share the program that facilitates the sharing of the data.

Below are some reasons why the virtual computer has become the best playground for programmers who want to create decentralized apps.

Very secure

Any item on the Ethereum Virtual Machine is encrypted and safe from outside hacks. The reason why the Ethereum Virtual Machine is so safe is that it is built on blockchain technology. This blockchain technology is very robust—not easy to destroy or bring to a stop, such as the Internet. It is the system other crypto-currencies are built on.

Transparent

The Ethereum network is like a public ledger, open for all to see. Everything that happens on the network is available to view. Despite this fact, the machine has a lot of privacy. A great example is with the smart contracts. You can quickly see a transaction that has occurred on the network, but the participants remain anonymous. That way, anyone can see how much Ether was transferred to your address, but they cannot tell that you are the owner of the said address.

The virtual computer also ensures that every piece of information is open to anyone on the network. The individual computers are each left with a local copy of the data. Thus, no one can claim to have lost a document or tamper with the original. The system cannot be censored.

The machine is unstoppable. The reason why it cannot be censored is that every individual computer on the network is both a node and a server. This ensures that there is no single point of vulnerability where a malicious hacker can take advantage of the system.

It would take a very ingenius hacker and a lot of human resources, as well, to bring down the network.

The decentralization of the Ethereum network ensures that it cannot be controlled by any one person or government. No one owns the network, nor can anyone censor it.

Interoperable by nature

The basic structure of Ethereum makes it interoperable. This means that data from one program can be shared and used by another application in the system. This is not possible on other platforms. This ability to share data and programs, however, is only possible if permission is granted.

Non-repudiable

This just means that the computer cannot be destroyed. This is because there is no single point of entry into the network. It also means that, if one machine should break down somewhere in the world, the system will keep on going. This is possible because the rest of the computers are working independently but are connected to the network and not a central mainframe. They are not sharing a single primary source of power or data storage.

Chapter 5: Understanding The Ethereum Infrastructure

Before you can begin investing in Ethereum, you need to know more about the infrastructure of the network. Like all network systems, there are quite a few moving parts that must be taken into consideration to understand fully how Ethereum actually works.

Smart Contracts

We've already talked a little about Smart Contracts. They are a means of adding code to the decentralized platform so that certain functions can be self-initiated without an additional party involved in the transaction. For this to happen, it was necessary to develop a system that could digitalize the terms of a contract so it could be stored in code on the network. So, rather than having a paper contract sitting in the cloud on a centralized server, instead, the parties would have a digitalized code that was stored on the blockchain.

According to the terms of the contract, the actions stipulated would automatically be executed once the specified parameters were met. Because Ethereum was created to adapt to a wide range of systems, it is possible to implement Smart Contracts on their blockchain. Of course, Ethereum is not the only system that has made these digitalized contracts possible, but to date, it is the most advanced.

It is important to understand that just because we use the term "contract" that were are not talking about contracts in the traditional understanding. It is actually a system of complex computer codes that allow the parties to exchange anything of value based on the terms and conditions stipulated. This way, users can exchange property, stocks, bonds, currency, or anything else that has an assigned value to it.

The system of Smart Contracts is much simpler than the traditional way of forming an agreement. Rather than obtaining a lawyer that will draw up the contract and witness signatures on the document these are self-executing. This saves an enormous amount of money in fees and hidden costs. With Smart Contracts, the fee is nominal and is paid in Ether and rather than waiting for weeks for your documents to arrive, the transaction is instant.

There are likely some ways of using Smart Contracts that you might not have considered yet, including:

Voting: With Smart Contracts, voting could be done entirely online. Once the terms of the contracts are determined, registered voters can cast their ballots during political elections without fear of the system being hacked or compromised in any way.

Real Estate: Real estate transactions can be easy with Ethereum. Once a sale is set up, it will remain there permanently with no chance of someone altering the terms.

Healthcare: Private healthcare records could be stored on the blockchain with access only permitted to those who have the codes set up in the Smart Contract. This could also be a valuable tool used when making insurance claims or other related costs.

There are likely hundreds more ways to use Smart Contracts in the future that go beyond the ability to trade currency. The more familiar you become with them, the more likely you'll find other ways to utilize them.

The Ethereum Virtual Machine (EVM)

The EVM is the key part of the system that allows for this adaptability. With this software program, users can create thousands of different applications that can run on one single platform eliminating the need to create a new Blockchain each time there is a need for something new. It is this EVM that makes it possible for Ethereum to adapt to so many different applications.

Decentralized Applications

Sometimes referred to as Dapps, these applications are created to fill a specific need. Each Dapp has its own function that it is designed to facilitate what the users expect. Just like the Blockchain itself, they are not maintained on a centralized server and can be used to perform all sorts of tasks.

Decentralized Autonomous Organizations (DAO)

DAOs are exactly what their name implies, organizations that are not managed by a single entity. These organizations work entirely on computer code and are dispersed among thousands of nodes within the Blockchain.

All of these parts work together to create a network that can be adapted to a wide range of applications. The benefits of this kind of Blockchain are endless. While there are several different elements to the Ethereum network, probably the most important is the Smart Contract. It is the single element that sets it apart from Bitcoin and many other digital currencies.

It is the Smart Contract that gives Ethereum the adaptability that makes it so appealing. Not only will you be able to create self-executing contracts and cut out the middleman entirely from future transactions, but it is also possible that there will be new applications formed in the future that Ethereum's founders could not possibly have imagined.

What you must remember is that Ethereum is not just a single entity but a compilation of several different aspects of the Internet.

Functions of Ethereum

Aside from the key elements of Ethereum, there are several tasks that the system must perform on a regular basis. You also need to understand these tasks and how they work before you begin to invest.

Mining

One way many people make money with Ethereum is through mining. This is a highly intensive computational work to find solutions to difficult math problems. The miner uses his computer's hardware along with a number of mining applications to verify transactions and create the "blocks" in the blockchain. To mine any type of cryptocurrency, it requires a great deal of time, computer space, and energy. The goal of the miner is to find the right "hash." which is done by generating a block and submitting it to the blockchain. Once it is verified, the miner is given a certain amount of Ether for his hard work.

Blocks & Transactions

No matter what you want to do with Ethereum, all operations are completed by means of transactions. Whether you want to transfer Ether from one account to another or you want to execute the transfer of data, it must be accomplished via a transaction. Once a transaction is entered into the system it is given a hash, a code used to obtain any necessary details related to it. Miners must confirm each of these hashes. Any transactions not confirmed are picked up, grouped, and analyzed to find the hash. Once the hash is found the block is generated and then added to the Blockchain.

Gas (Ether)

Any transactions performed on the blockchain are run by gas. This means that each transaction is assigned a gas limit and a fee. This gas is obtained from the miners when they execute the code. The total amount of fuel needed to process a transaction has to be less than the set gas limit. As a miner, you get to collect the fee for every transaction you confirm.

Transaction Costs (Fees)

The fee for each transaction will depend on how much gas it took to confirm the transaction. For example, if the cost of adding two numbers was 3 gas then the formula used would look something like this:

Total cost = gas used x gas price

All miners use a default strategy like this one to determine how much of a fee they should charge. If the amount the user is willing to pay is less, then the miner will reject the transaction. Obviously, the gas price will vary depending on how much is involved in confirming a transaction.

Swarm

Swarm is basically the storage platform used to hold all the data uploaded on Ethereum. It is completely decentralized and holds and distributes all of the Dapp codes used on the blockchain. It works the same way as the world-wide-web except that the data is not contained on a specific server and is spread across thousands of computers around the globe. This allows for zero downtime and a powerful resistance to security breaches.

Whisper

Whisper is a built-in protocol that allows the different Dapps to communicate with each other in order create a transaction.

Ethereum Naming Service (ENS)

The ENS is a special service that is capable of renaming human-readable addresses into codes that can be read by machines.

Consensus Algorithm

The consensus algorithm is a protocol used to keep all transactions on a blockchain in a particular order. It is the means by which the network can maintain its security by acting as a deterrent to denial of service attacks and other common abuses often found on the Internet. Many different algorithms can be used, but the two most common ones are Proof of Work (PoW) and the Proof of Stake (PoS)

Proof of Work: The PoW allows users to know if the chain is valid because the miner's work is shown. With PoW, it is not necessary for every node in the system to submit their conclusions for every transaction. Instead, it utilizes a 'hash function' that sets up certain conditions where a single miner is allowed to present his or her findings, which can then be verified by other participants in the system. Bitcoin uses the PoW algorithm in verifying transactions before they are added to the blockchain.

Proof of Stake: With PoS, miners are required to put up or stake a certain amount of their own coins to verify a block on the blockchain. Basically, they are investing their own money in the network. When it comes to making calculations, PoS calculations are much easier to solve. The miners need only show that they are the owners of a given percentage of the coins in that currency. So, if they can prove that they own 5% of the Ether, then they would be allowed to mine 5% of any transactions made on the network.

As you can see, there is a lot involved in the Ethereum network. There is no question that it is an innovative technology that can transform the Blockchain in many ways. Of course, we are only able to touch on the inner workings of the system, but we hope it is enough to whet your appetite to learn more. If you're seriously planning on making money in Ethereum, it is very important that you delve even deeper into the details of this system. In the meantime, it should be enough for you to get a basic understanding of the system before you start investing your money.

Chapter 6: The Technology Behind
Ethereum

The Ethereum network allows developers to build and use decentralized applications (dApps) that serve specific purposes to its users. For example, Bitcoin is a form of dApp that provides users with a peer-to-peer electronic cash system that allows online Bitcoin payments. DApps are composed of code that are running on a blockchain network and not controlled by any central entity or individual.

With Ethereum, any centralized services could be decentralized. Consider all intermediary services, which exist across different industries – from conventional services (such as loans offered by financial institutions) to intermediary services (like regulatory compliance, voting systems, and title registries).

The Ethereum network can also be used in building decentralized autonomous organizations (DAOs), which are completely independent and have no centralized authority. These are operated by programming code, on a collection of smart contracts integrated in the Ethereum blockchain. The code is programmed to alter the structure and rules of a conventional organization to get rid of the need for centralized regulation. A DAO is owned by everyone who buy tokens, but tokens are equal to ownership and equity shares. The tokens serve as a symbol of contribution that grants people with voting rights.

Ethereum Decentralized Platform - The Pros

Since dApps are running using blockchain technology, they can also take advantage of its beneficial properties. This includes:

• Zero downtime – applications never go down and could never be switched off.

• Security – With no centralized point of failure and improved security through cryptography, decentralized applications are well protected against fraudulent activities and hack attempts.

• Tamper and corruption proof – Applications are based on a network that is built around the concept of consensus, making censorship impossible.

• Immutability — no third party can alter the data in the application

Ethereum Decentralized Platform - The Cons

Even though Ethereum offers numerous advantages, it's not 100% foolproof. Remember, the code for smart contracts is written by humans, so they are only as good as the programmers. Oversights or code bugs could lead to unintended adverse effects. If a bug in the code is taken advantage, there is no surefire way of preventing the attack other than getting a network consensus and then reprogramming the root code. This is against the core concept of blockchain, given that it's designed to be immutable. In addition, any action that is taken by a centralized party could raise serious questions about the decentralized nature of an application.

Developing a Decentralized Application in the Ethereum Network

There are several ways that you could plug into the Ethereum platform. Among the easiest ways is using the native Mist browser that provides a basic interface and online wallet for storing and trading Ether and to create, manage, and store smart contracts. Similar to browsers that provide access and help people to browse the internet, Mist offers a portal into the world of dApps.

Another extension is MetaMask, which converts Google Chrome into the Ethereum platform. MetaMask enables anyone to easily develop and run dApps from their browser.While it is originally designed as a Chrome plugin, MetaMask will gradually be available for Firefox and a range of other web browsers.

Although these are still on beta version, MetaMask, Mist, and other browsers are set to make blockchain applications more accessible in the coming years. Even users without technical background could easily build blockchain applications, which is a significant development for blockchain technology and could raise dApps into mainstream use.

Current Applications Being Developed in the Ethereum Network

People can use the Ethereum network to develop applications across a wide range of industries and services. However, developers are exploring a new field, so it can be difficult to figure out the applications that will succeed or fail. Below are some interesting applications that are presently in the Ethereum network:

• Augur – This open-source application is designed to predict events and receive rewards for correct forecasts. Predictions on future real world events such as who will become the next US President, are implemented through virtual shares. When a user purchases shares in a winning prediction, they can receive monetary rewards.

• Provenance – This application harnesses the power of the Ethereum network to improve the transparency of opaque supply chains. By keeping the origins and record of products, the project can build an accessible and open framework of information so consumers could make informed decisions when they purchase products.

- BlockApps – Provides an easy platform for businesses to build, manage, and use blockchain applications. From integration with legacy systems to full production and proof of concept, this app offers all the tools needed to develop private, semi-private, and public industry-specific blockchain apps.

- Uport – This application provides users with a convenient and secure way to take charge of personal information online. Users can control who can access and use their personal information instead of depending on government agencies or entrusting their personal data to third parties.

- Weifund – This open application provides an easy platform for crowdfunding campaigns that harness the power of smart contracts. It allows contributions to be converted into contractual-based digital assets, which could be used or traded inside the Ethereum platform.

The 2016 DAO Attack

At this point, you should already know the fact that the Ethereum network can be used to build DAOs. Well, in 2016, a DAO has been compromised due to a hacking attack. A startup company developed a DAO to provide a humanless venture capital service, which aims to help investors make decisions through the aid of smart contracts.The DAO was financed via an ICO and ended up raising about $150 Million from investors around the world.

Upon raising funds, The DAO software was hacked by cyber criminals who took Ether worth around $50 Million during that time. Although the negligence was not on the side of the Ethereum platform but because of the technical flaw in the system of DAO software, the executives of Ethereum were obliged to address the mess created by the attack.

After all the discussions, the Ethereum community came to an agreement to recover $50 million dollars' worth of Ether by making a change in the code – a process also known as a hard fork. The change in code moved the robbed funds to a fresh smart contract created to allow the original owners to draw back their tokens. However, the ramification of this decision is still controversial and debatable.

Ethereum is based on the blockchain network wherein every action recorded is irrevocable and permanent. By changing a code and writing the rules by which blockchains operate, Ethereum made a risky precedent that contradicts the very purpose of blockchain. If blockchain is modified every time a huge amount of money is involved or a number of users get affected, the major value proposition of blockchain (secure, anonymous, tamper-proof, and unchangeable) will be lost

While a fork was put forth, the Ethereum community and its executives were trapped in a perilous situation. If they failed to recover the robbed fund of the investors, trust in Ethereum could be lost. On the other side, retrieving the fund of the investors set actions that go against the main essence of decentralization and require a risky precedent.

At end of the day, the Ethereum Community came to a consensus to execute the hard fork and recover the stolen funds of DAO investors. But not all agreed with their decision. This lead to a division, creating the two parallel blockchains that now exist.

Ethereum Classic (ETC) is available for those who went against the hard fork.Ethereum (ETH) is available for most members of the network who voted yes to change a minor part of the blockchain and issue a refund to the rightful owners.

Ethereum Classic and Ethereum both have similar features and are essentially the same in every way, up to a certain part where the blockchain was rewritten or changed. This only shows that all that occurred on Ethereum up until the change in code is still valid on the Ethereum

Classic blockchain. From the part where the hard fork was performed onwards, the two blockchains work separately.

Despite the controversies from the DAO software hack, Ethereum is still looking forward to big opportunities. By giving a user-friendly platform that allows users to equip the functions of blockchain technology, it is now helping in quickening the decentralization of the world economy. Apps that went through decentralization have the possibility to completely disrupt entire industries such as insurance, academia, real estate, health care, finance, and many more.

Chapter 7: Effect Of Blockchain On Computing

Blockchain computing is the most talked term in the IT industry because it is secure and cost-efficient for everyone. It is important to understand its features, architecture and model:

Features of Blockchain Computing

There are five prominent features of blockchain computing that will help you to understand its efficiency:

On-demand Service

You can use the blockchain to configure and deploy apps without any heavy lifting of information technology. Some vendors offer a template to configure your work because it is based on self-service and service models available on-demand. It will help users to interact with the blockchain and perform different tasks, such as deploying, building, scheduling and management. It will help you to access computing capabilities and help users to bring suppleness to their work for current and future uses.

Resource Pooling

Blockchain will enable you to centralize your IT resources and spread its use to available servers. This can maximize the power of shared computing to distribute the capacity as per your requirements. You can get the advantage of peak-load capacity and utilization efficiency. Allocation of resources will be elastic and you can change it to demand.

Virtualization

It is an important feature of the blockchain because you can create a virtual version of your application topologies and move these topologies across the blockchains and between data centers. Blockchain can increase the accountability of scalability and use. It will make electronic communication easy with the help of networks and devices. The vendors work to provide access to network, storage, memory and CPU.

Accessibility

Blockchain enables your business to launch applications on different platforms, such as Apple TV, android phones, and laptops. The resources will be accessible and reliable. You can use the blockchain on a tablet in the absence of your office network.

Scalability

Blockchain has the ability to scale up and down without hoarding data. Computing capacity may demand spikes and on-demand scalability is similar to elasticity.

Fundamental Concepts of Blockchain Computing

Blockchain refers to a discrete IT environment designed for remote provisioning and gauges IT resources. This term is designed as a metaphor for the internet and offers to remove access to decentralized IT resources. Initially, the symbol of the blockchain was used as a representation of the internet in a variety of specifications and conventional documentation for web-based architecture. Blockchain environments offer IT resources and supply back-end processing competencies. It is based on internet protocols and technologies. Protocols are standard methods used to promote easy communication in a pre-defined manner.

Management and Security Mechanisms

Blockchain computing may pose privacy concerns because the service provider has an ability to access your data in the blockchain anytime. The blockchain can be altered deliberately or accidently or it can even delete information. Some blockchain providers often share details with third parties for the purpose of security and law and order devoid of the warrant. It is written in their privacy policy and the users have to agree on this policy before using services of the blockchain. You will own the legal ownership of data and you can get physical control of equipment because it is more secure. The end users may not understand the risks and issues involved in the blockchain services because they often click "Accept" without even reading it.

Blockchain Architecture

Systems architecture involves the delivery of blockchain computing with multiple components with each other on the loose coupling mechanism, including messaging queue. The elastic provision involves intelligence in the use of tight coupling or loose coupling to mechanisms.

Blockchain Engineering

It is an application of engineering disciplines to blockchain computing and bring a systematic approach to standardization, commercialization and supremacy in operating, conceiving, maintenance and development. It is a multidisciplinary method surrounding contributions from different areas, such as information, performance, web, security, platform, risk, quality engineering, and software.

Types of Blockchain

Blockchain computing is basically based on resource sharing than handling applications via individual devices and local servers. It will be good to use the internet enabled devices for storage and other services. Blockchain will allow the functions of different apps, such as virtual servers and desktop applications. It is easy to take the advantage of blockchain computing and resource sharing the blockchain computing is based on different deployment models.

Public Blockchain

It is a blockchain hosting service delivered to the network for public use and this model can represent blockchain hosting in a better way. The service providers often render infrastructure and services to various clients. Customers can have control on the location of infrastructure and from a technical viewpoint; there is no different between public and private blockchain structure except the level of security offered for numerous services given to the blockchain subscribers and hosting providers.

It is suitable for the business requirement to manage the load and host Saas-based application and many users can consume it. The operational cost of this model is economical and the deal often provides free services or in the form of a particular license policy. The cost can be shared by all users and the public blockchain is profitable for customers by achieving a financial system of scale. The public can get the advantage of free blockchain and it is "Google".

Private Blockchain

It is internal blockchain and the platform is implemented in the blockchain-based environment that is secure and often safeguarded by the firewall. It remains under the governance of IT section of any corporation. This blockchain allows only authorized users and offers a great control to organizations on data. It is tricky to define constitutes of any private blockchain. It may be difficult to define it according to services and variations. The physical systems are hosted internally and externally and they offer resources from a discrete pool for the private services of the blockchain. Business organizations often require these services for private alarm, assignment, and uptime requirements. Security may be evaded in the private blockchain in case of natural catastrophe and interior data theft.

Hybrid Blockchain

It is an important type of blockchain computing and it can be arranged for two or more servers. You can get the advantage of multiple deployment models. It can cross isolation and overcome limitations of the provider. It can't be categorized into private and public community blockchain. It enables users to increase the capability by assimilation, aggregation, and customization with blockchain package and service. The resources can be managed by external or in-house providers. It can be an adaptation among two platforms to exchange the workloads between public and private blockchains as per your needs.

Non-critical resources, such as development and workload testing can be accommodated in the public blockchain and it may belong to a third-party provider. An e-commerce website may be hosted on a private blockchain and offer security and scalability. It is not a major concern for the site of the brochure and hosted on a public blockchain will be really cheap as compared to the private blockchain. The businesses will be more focused on demand and security of their unique presence and implement an effective strategy. It will make your business focused and their unique presence is often implemented on a hybrid blockchain as an efficient strategy. It is known as blockchain bursting and it can be accessed with the hybrid blockchain.

Organizations often use this model to process big data. The private blockchain encompasses business, sales, and various data. It will initiate different queries on the public blockchain because it is efficient to meet the growing demands. Hybrid has flexibility, scalability, and security. You should learn to deal with network connectivity issues and expenditures because you have to deal with a few challenges while using this.

Community Blockchain

It is a good type of hosting and it helps you to setup mutually shared software for your organization, such as bank and trading organization. It will be a multi-talent setup for several organizations and the community members often share same performance, privacy along with security concerns. The basic intention of this software is to achieve business objectives and the blockchain can internally manage by the third party provider. The cost will be shared by organizations in the similar community. It is possible to use this computing service to manage, implement and build similar projects. The organizations can understand its potential and they often require this service for their business.

Chapter 8: Mining & Investing, Let's

Break It Down

We know how to buy cryptocurrency, we know how to sell it, we know about exchanges, and we've covered how to do it all, but now we need to get onto the subject of investing. Before we head there however, we need to touch upon another subject – mining.

Now, we're not going to go into the hugest amount of detail here, because unless you are someone who wants to really get into the serious depths of cryptocurrency at the very start, you're a hacker, or you're someone who actually understands algorithms, then it's best to just know what it is, and how it is pertinent to cryptocurrency as a whole at this time. As you move onwards towards your master journey however, mining is definitely something you could look into, as you gain more knowledge, confidence, and experience.

Mining? No Tin Hat Required ...

Mining has nothing to do with donning a protective hat and heading underground, but it has a lot to do with using your brain and using sophisticated technology. First things first however, we need to talk about what mining actually is.

We know that when a transaction is requested to take place, it has to be encrypted and verified, in order to create another block in the block chain. Because cryptocurrency is decentralised, it relies upon the block chain to record everything, and ensure that there is some amount of control, e.g. no duplicate spending. Mining is the process of the computer actually verifying a transaction and writing that next block on a block chain. By doing this, you are mining coins, and creating more. It stands to reason then that the more miners a cryptocurrency has, the stronger it is and the better is performing.

So, to break it down to its simplest definition – mining is the process of actually creating cryptocurrency.

You can become a miner, hat optional. It is not the best idea to sign up to be a miner for Bitcoin or Ethereum, because their mining pool is so hugely over-subscribed, that you're simply not going to get a look in. You could however start your mining route with a lower Altcoin, or if you really want to get further into the cryptocurrency world, then you could start your own cryptocurrency, and mine it yourself. A little more on that later.

To give you a quick overview of what you will need if you do want to move into the mining world a little later on in your journey:

•You will need the resources to actually be able to carry out the mining process. This is a strong Internet connection, without disruption, and a good quality desktop or laptop computer. You will be on your computer a lot if you do get into mining, so it needs to be strong, and it's a good idea to make sure the fan is in good working order, to avoid overheating and crashing. As a miner, a crash really would be a disaster.

•You will need to download the mining software, which is available online. Do some research into which is the best for you, according to the cryptocurrency you're planning on mining.

●You will then need to register for the mining pool of that particular cryptocurrency, and complete your application.

A miner is 'paid' or rewarded with a miner's reward, and this is basically cryptocurrency back in your wallet. So if you did end up mining for Bitcoin, you would be rewarded with Bitcoins for your troubles.

How to Invest in Cryptocurrency

In order to know how to invest in cryptocurrency, we first need to understand the subtle difference between trading and investing. We've already talked in depth about trading, in our last chapter, and whilst investing may sound very similar indeed, the overall aim is a little different.

Trading is buying cryptocurrency and then selling it on for a profit, before buying more and repeating the process. This is a quick form of investment, which is constantly moving, and never stands still for too long. Trading is something which takes up a lot of time and is constantly moving.

Investment on the other hand is a more long-term kind of deal. Whilst you do need to be very switched on and monitoring the situation, investment is perfect for busy people, because it doesn't take up too much time. You simply invest in the right cryptocurrency for you, and you wait until the perfect moment to sell it on. You are basically buying, and holding, until the time is right.

Those are the key differences between trading and investment in its purest form.

Now, how to actually invest in cryptocurrency is actually very similar to trading, ironically! It's just the time frame involved which is different. Because investing is spread over a longer period of time, there are certain personality traits and qualities that an investor of any kind needs:

- Patience
- Determination

- The ability to analyse and monitor, without panicking
- Nerves of steel

If you invest in Bitcoin you are going to notice that the rate of value fluctuates wildly across the space of a day. An inexperienced investor will panic about this and perhaps trade in prematurely. What you need to realise is that it is totally normal for this to happen, because Bitcoin by its general definition is quite volatile. The lesser Altcoins don't fluctuate anywhere near as much, but they also don't bring in the same large gains that Bitcoins do. This is where you need to weigh up the pros and cons of which cryptocurrency to actually pile your time, cash, and attention into.

The process of investing will sound quite familiar to you by now:

- Research the right type of cryptocurrency for you to invest in, weighing up the potential gains, its market performance, etc
- Decide how much you want to invest

- Purchase the cryptocurrency of your choice through an exchange, using one of the three methods we mentioned (broker, direct trading, or a trading platform)
- Wait
- Wait
- Wait some more
- Whilst you're waiting, watch the market and watch the performance of your particular cryptocurrency
- Read projections on the cryptocurrency
- When the time is right, when you are sure you have reached a point of maximum gain, cash in your investment
- You can then run, or you can reinvest somewhere else

This example shows the similarities between cryptocurrency investment and trading in practice.

Now, the secret of being a good investor is knowing which investment to take on, and knowing when to cut your losses and cash in.

How to Choose The Best Investment, & When is The Best Time to Cash in?

What does a good investment look like?

This is a difficult question to answer, because a bad investment can be a wolf in sheep's clothing, and what seems like a good investment could quickly go south. Basically, it's about weighing up the risk versus benefit, and actually identifying the amount of risk involved.

As new cryptocurrencies are brought onto the market, their value is low, but the more miners that work on the currency, the stronger it gets, and the more growth occurs. If you choose carefully, and make an investment just before a boom, you can make a serious profit. Of course, these Altcoins have a tendency to boom quite dramatically, and then fall just as dramatically and just as quickly. When investing in an Altcoin that isn't Bitcoin or Ethereum, it's best to cash in on your investment as the peak comes, or just the second before it. The value will drop sharply afterwards, and even if not, it will certainly lose its value over a shorter period of time than you may expect.

Of course, real investing rarely consists of just one investment, but usually a few.

When you're investing in more than one cryptocurrency, to spread your potential gain, you will needseveral wallets on the go at any one time. That can be confusing, so how do you keep track of everything?

How to Monitor Your Investment

You've made an investment, you've assessed the risk prior, but how do you monitor it? Where do you gain the information, in order to know when to cash in before a crash occurs?

There are a few ways to do this.

Of course, you're going to watch the market closely, and you can do this online, and by downloading specific apps and software. You don't have to watch an investment every second of the day; investing isn't as heavy going time-wise as trading is, but you do need to regularly check, to ensure you're not missing out on a rapid upturn, or downturn. We also just mentioned the fact that you have several wallets working at any one time – can you keep them all in one place?

Downloading a cryptocurrency comparison tool is the best option here. Not only will this give you all the information you need in terms of performance against each other, but you can hold your wallets in one simple and easy to use estimation, rather than having them all over the place. When you create your account you will need to share information in terms of your addresses of your various cryptocurrencies, and this will need to be verified, so it can be a time consuming process. It is worth it however, and the best way to find the most high quality comparison tool is by checking online reviews. There are many out there, so do take the time to really check this out carefully. The benefit will outweigh the time taken.

Once done, you can check performance and potential gain, as well as having access to all your wallets across one portfolio. Easy!

Trading in The Short Term & Longer Term Investing

We've talked about how to invest, and by this point in the book, the fog of cryptocurrency should definitely be clearing. One question you might be sure of however is whether short term trading is best for you, or whether you should go for longer term investing.

Let's quickly check out the pros and cons of each.

Short Term Trading Pros

●Ideal for those who want to make a smaller profit, faster

●A great way to learn more about cryptocurrency investing, before making a larger investment

●Generally less risk involved, as lesser amounts of cash

●Short term trading can turn out to be a very lucrative hobby

Short Term Trading Cons

●Smaller profit margins involved

●It is time consuming, and may not be ideal for those who lead busy lives

●The more frequent use of an exchange may mean more fees which build up

●Finding a good exchange is even more important, due to possible restrictions in place

Longer Term Investment Pros

• The end financial gain is much greater

• Less time consuming than trading, which is much more intensive

• An investment comparison tool can be downloaded, to enable comparisons to be made, and keep everything in one place and to make monitoring easy

• There is less chance of a sudden dip with longer term investments, compared to short term trading

Longer Term Investment Cons

• The gain doesn't come quickly, so you have to wait for your cash!

• Knowing when to cash in, and when to wait is something which needs to be learned over time

• Choosing the initial investment can be tricky

From that list of pros and cons of each avenue, you can see which one will be best for you. It really does come down to time and how intensive it is. It depends how much of your life you want to dedicate at this stage to cryptocurrency, and how hands on you want to be initially. Of course, it also depends on whether you want small profits regularly, or larger profits over a longer period of time. Many people view this as swings and roundabouts, but it depends on your personal circumstances as to which one you should choose. Of course, you could do both, but this is going to eat into your time quite considerably!

Investment is certainly something which needs to be weighed up carefully and it is something which requires careful consideration before the cash is transferred to the investment itself. Taking on a small investment first will help your confidence grow slowly and from there you can build up to bigger investments over time.

Chapter 9: How To Reduce Your Risks With These Investments

As a new investor, you want to make sure that you are decreasing your risks as much as possible. Any investment is going to have some risks, but if you jump into the investment without doing your research or coming up with a strategy, you are more likely to fail. There is a reason that some people make a lot of money in cryptocurrency and others fail quickly. This chapter is going to take a look at some of the easy things that you can do to help decrease your risks while working with Bitcoin and Ethereum.

Diversify your portfolio

One of the best things that you can do no matter what kind of investment you choose, is to diversify your portfolio. As a beginner, you may be excited about getting into the market and making some of your first investments. You want to jump into it all with two feet, but there are some issues that can come with this. You may be busy thinking about all the profit you can make when you put all your money into one investment. But what most investors don't think about is what happens if that investment does poorly.

If you put all of your money into one investment, you are increasing your risks quite a bit, even in an investment that is relatively safe. If that investment starts to do poorly, you will lose out on all your money and will have to restart from square one, which can be hard on many new beginners. Any investment can take a turn for the worse at times, many times due to factors that you can't control so it is best to plan for this.

To avoid this from happening, it is best to diversify your portfolio. What this means is that you take the money that you would like to invest, and put some of it into a few different investments. A number of investments that you can split your money between will depend on how much you have available, but the more times you split it up, the safer your money is.

The nice thing about this option is that even if one stock does poorly, you will only lose a little bit of money instead of all your money. And it is likely that at least one or more of your other options will increase in profit. You are much more likely to see a nice profit from your options when you split it up a bit, rather than putting all your money in one place.

Invest in what you know

Getting started in Bitcoin and Ethereum can be tough. It is a new adventure and you want to make sure that you are doing it the right way, but there are so many options and you may not be all that familiar with it. The best thing that you can remember is to invest in something that you know about.

Both of these currencies have a lot of options for you to choose from so there is sure to be something that you are knowledgeable about. Perhaps take some time to look through the Ethereum platform and see which apps are about things you know. While there are a lot of great apps and programs that will become available, you need to make sure that you are picking out something that you have some idea about. You can choose an industry that you are familiar with if that helps. When you know about the industry or the product, you are better able to tell if it is the best option for you to choose.

This can happen with Bitcoin as well. You want to make sure that you don't jump into something that you are completely confused about no matter which option you choose. Let's say that you like to go hunting and fishing. There may be an app or a company that sells these products and you can invest in them. If you are knowledgeable about another type of industry or company, consider going with those.

It is tempting to jump into a company or another investment because it looks like it is doing well, but if you know nothing about that industry or that investment, you are going to run into some trouble. You won't be able to pay attention to the news or what is going on around you, and you will end up doing poorly in the long run. Stick with what you know; even in the new cryptocurrency market, you are sure to find something that you know about.

Keep your emotions out of the mix

One of the most important things that you can do when working with Bitcoin or Ethereum is to keep your emotions out of the game. These two options are brand new investments, which can be both exciting and scary. It is scary because there isn't a lot of information about these investment types since it is so new. But there is also some excitement since there are so many possibilities that can come with this option.

It can be hard to keep your emotions out of the game though, especially for someone who is new. There are times when you may put your money into the investment choice and then the market takes a downward turn right away. Usually, these are going to be pretty small and if you keep your money in for a little bit longer, you will end up making all that money back and more. But most people get a bit scared and freaked out, and then they withdraw their money. This results in a loss that could have easily turned into a gain if they didn't let their emotions get in the way.

This can happen to go the other way as well. Some people get so excited when they see that their investment is making money. They see it go higher and higher and they assume those highs will just keep on coming. But at some point, the investment will go down again to reach that mean. And if you hold onto your asset for too long you will see all those potential profits go down the drain. Once the market turns, it turns quickly and you can quickly lose out on everything because you got too excited.

Keeping your emotions out of the mix is the best thing that you can do. Make sure that you set up a strategy that tells you when to enter the market and when to leave will help to take some of the emotions out. This helps you to make smart decisions and then when one of the points of your strategy is met, whether it is the high point or the low point, you will leave and reduce your risk. Never let those emotions get in the way of you making good money on your investments.

Do your research

Since Ethereum and Bitcoin have not been around for a long time, the amount of research that you can do about their investment options is a bit more limited compared to other options on the stock market. But this doesn't mean that you shouldn't do your research ahead of time. There is always going to be some research, no matter how recent the company is, that you can use to get the results that you want.

With Bitcoin, many people are choosing to just purchase the digital crypto coins and hold onto it to make a profit later when the prices rise. But there is still a good amount of research that you can do. You need to have a good idea of how much Bitcoin actually costs, how much it will rise in value in the future compared to your method of paying out (how much the Bitcoin is worth per American Dollar, for instance), and how much you should purchase. Researching the trends in Bitcoin and its value in different currencies can help you to make this decision.

Ethereum is a bit different and you will often work with apps and other startups instead of just investing in the currency, although that is an option as well. Get as much information about these apps and companies before starting. You want to make sure that they are going to grow their business well and won't just tank or take your money and run. If you are uncertain about where to start, talk to someone who has worked in these investment options and ask lots of questions to help you out.

Keep some savings on hand

No matter which of these two options you choose to go with, it is important that you always keep some savings on hand. When you put all of your money into the investment, you could end up in trouble if the market goes down. Having some savings on hand, especially savings that is in cash, can help you out because it ensures that you are set in case something does go wrong.

Many new investors don't think about this step, and so they will invest all of the money that they can, not thinking about what will happen to them if all that money disappears. It is best to choose a comfortable amount that you would like to invest and stick with that. You may not make as much profit as you could this way, but you at least won't lose as much as you could either.

Never invest money you don't have

This is a big beginner's mistake that you have to avoid. It is tempting to look at some of the numbers that have come on the market with Bitcoin and Ethereum and to just throw all your money right at it. But no matter how well these cryptocurrencies have done, they still have some dips and if you don't invest wisely, you could end up losing your money.

The best method for investing is to only invest the amount that you are comfortable and willing to lose. Don't take out your retirement, don't use all your life savings, and don't take out a crazy loan to pay for this investment. Only use money that if you lost, you would still be fine afterward. It is tempting to do a big investment to see a big return in the process, but the market can easily turn the other way and you could lose out on everything.

Know when to give in

No matter which way the market is going, it is important for you to know when it is time to give up. Actually this could be very tough to learn. There are times when you need to ride the market, even when it goes down a bit, but then there are times when you need to cut your losses and then move on to your next choice. Even when the market is going up, too. Yes, you may see that the market for that investment keeps creeping up, but at some point, the market will go back to the middle and if you don't learn the right time to get out, you will end up losing money instead of gaining.

One of the best ways to make sure that you don't go too far in either direction is to set an exit point. This is the point when you will get out of the market, no matter what it might do when you are done. You can set a low and a high exit point. So when the investment hits your low point, you will cut your losses and call it good. But when the investment hits the high point, you will withdraw and keep your earnings. There will be times when you miss out on some profit when the market goes above your exit point, but it is a lot better than losing out.

The point of doing this is to help you take your emotions out. As soon as those emotions get into the mix, you are going to take things too far and end up losing money. Set the points that you are comfortable with right from the beginning. If Bitcoin or Ether(eum) ends up going higher, you can always enter the market later on, but this will prevent you from losing too much money.

While any type of investment is going to be a risk, there are steps that you can take to minimize these risks a little bit. Learning how to mix up your portfolio a bit so you have different options (in case one doesn't work out that well for you), taking the emotions out of the game, and coming up with a good investment strategy is one of the best ways for you to see success.

Chapter 10: Mining Ethereum

If you are interested in mining Ethereum you are going to need to understand a little bit more about the process and what exactly you are going to be doing. Ethereum mining is currently completed using a specifically-built mining machine that uses a SHA256 double round hash verification model as a means of verifying blocks before they are added to the Ethereum blockchain directly so that it is clear that all of the related transactions occurred in the fashion that the block says they did. The speed at which a machine can do this is generally written as hashes per second.

In exchange for your time and energy costs, each block that you successfully mine will reward you with 5 ether, along with a portion of the gas fee that was paid by the person who put the transaction in motion in the first place. This amount should be more than enough to compensate for the costs of the electricity required to mine the block and still leave enough left over to turn a profit. The greater the processing power of your mining machine, the faster you will be able to verify the blocks and the more you will make overall.

This level of verification is required to ensure that things continue to run smoothly as the Ethereum blockchain is decentralized which means that there is no guiding hand making sure that everything is working properly. Rather, everything runs via automated processes to keep things working properly, as long as the transactions are verified by a third-party source. The proof of work that is required to do so is difficult to produce and occurs via what is essentially a randomized process which is why the high-powered mining machines are required in order to ensure that the right answer is found as quickly as possible.

As previously mentioned, the most common proof of work that is used by many cryptocurrencies is the SHA256 hashcash. This proof of work model was first created in the 1980s as a means of blocking spammers from sending out endless emails through the emerging email technology. The proof of work required the computer sending the email to complete an equation for each email sent, with the proof getting more complicated the more emails that were sent. This would have been enough to stop 1980s computers in their tracks, but it never saw widespread adoption.

It gained new life as part of the blockchain paradigm, however, and it can be tweaked in such a way that its difficulty will ensure that blocks are not verified faster than they can be generated by the chain. Due to the random elements inherent in the process, it is impossible to determine who is going to be eligible to mine the next block ahead of time. If a new block is to be considered validated, its hash rate is going to need to be less than the block that proceeds it, which naturally leads to blocks to self-sort themselves.

When added to the chain, each block is then given its own hash function that corresponds in a random way to the transactions it contains. This means that if the block is ever tampered with, the hash function will change and the blockchain will naturally notice the error. The only way this would not be the case is if every block relating to a specific transaction was changed at the same time.

Start mining

In order to get started mining, the first thing you will need to do is find out the current state of mining technology. The best way to do so is to check out the Ethereum subreddit to get a feel for the current power and range of prices available on the market.

While you could theoretically get started mining Ethereum using the CPU of your laptop or the graphics card in your PC, doing so is akin to racing in a Formula-1 race with a stock Honda Civic, you can certainly do it but you likely aren't going to get very far. Mining machines are specialized in such a way that they can finish proof of work requirements up to a hundred times faster than the average computer. This means that if you were to get started without the right equipment then you would still likely be waiting to finish your first block six months from now. It should go without saying that this means it would cost you far more in electricity then you could ever hope to make back in profit. Prebuilt mining machines tend to run between $500 and $4,000.

Once you have a machine, the next thing you are going to need to do is download the program that will be used to actually mine the Ethereum blockchain. There are numerous different types of this software but the two most popular are going to be BFGMiner and CGMiner, they both run from a command line, however, so if you prefer something with a more modern interface then you are going to want to go with EasyMiner instead. EasyMiner is available for Linux, Windows and Android devices.

Once you are ready to get started, the last thing you are going to need to do is choose a mining pool to join. Mining pools are simply loose associations of miners that come together to pool the power of their mining machines in order to generate proofs for blocks as quickly as possible. While it is technically not required for you to join a mining pool in order to mine Ethereum, it is the most profitable way to go about doing so as the difficulty of the average proof continues to become more complicated the more transactions the blockchain holds overall. While you will need to split the profits from the block with the rest of the pool, you will find that the speed at which you finish each block is going to more than make up for the difference.

If you decide to not use a mining pool, then you will want to download the official Ethereum core at Ethereum.org which will ensure that your machine remains in-sync with the blockchain at all times. If you are planning on joining a pool, however, then all you will need to do is to review the requirements and conduct for the pool and ensure you follow the rules.

In order to ensure that you can be paid for your work, you are going to need a home for your ether which means finding an ether wallet that works for you. Choosing the right wallet is a big responsibility as it is going to be the last thing standing between your ether and outright theft. There are two main types of wallets, hot wallets and cold wallets, with hot wallets referring to those that are connected to the internet and cold wallets referring to those that are not.

Hot wallets can be either software based on website based. Software wallets are safer than online wallets as they are stored on a device directly as opposed to on a server. Despite the name, an ether wallet won't actually hold your ether, as that is stored immutably on the Ethereum blockchain. Instead, what it is storing is your access to your ether through what is known as a private key. A private key, is just that, it is the secret access code to your ether which is why you need to keep it safe as whoever has it can do with your ether what they will. You will also have a public key which you can give out to have other people send you ether. Software wallets can be used to complete transactions regardless of whether or not your device is currently connected to the internet.

Cold wallets come in several forms, the first of which is a hardware wallet. A hardware wallet is typically a USB storage device that is encrypted in such a way that it is impossible for someone to remove your private key from it in text file format. Many hardware wallets also have screens that allow you to interact with your cryptocurrency directly, without having to plug it into a computer and thus potentially compromise its security.

Even more secure than a hardware wallet is a paper wallet. With a paper wallet, you use a website that you download to your computer to generate an address for a wallet while the computer is not connected to the internet. You then print off the QR codes that are generated from the process and store the one that is private while snapping a picture of the other with your phone and using it to complete day to day transactions. When done correctly, this is one of the most secure ways to store your ether as possible as someone would have to physically find the piece of paper with your private key on it in order to steal it. Even better, the entire process is completely free. You can find complete directions for doing so at WalletGenerator.net.

While a paper wallet is almost completely secure because it only requires you to take care of a single sheet of paper, a brain wallet is 100 percent secure, assuming you have a good enough memory to entrust it with the full amount of your ether wallet. A brain wallet is similar to a paper wallet, in that you create one by accessing a saved version of a website from an offline computer. What is actually happening with that website, however, is quite different. When you create a paper wallet you are generating a permanent link to your private key in the form of a QR code. With a brain wallet, you first choose a phrase, 12 words or longer is recommended, that other people are unlikely be able to guess by plugging 12 random words into the system. The words then create a link to a specific wallet which can be generated any time you need to interact with your ether. When you aren't doing so there is nothing to link you to the wallet in question besides a random phrase. You can find complete directions for setting up a brain wallet at BitAddress.org.

Build your own mining machine

To ensure you can begin mining ether successfully with as little money down as possible, you are going to want to build your own mining machine rather than purchasing a premade alternative. Doing so will allow you to get somewhere between two and a half or three times the power as what you would get for the same cost in a premade machine.

Graphics cards: In order to ensure you are generating enough processing power for the mining process you are going to want to use Graphical Processing Units (GPUs) which are commonly found in video cards that are often used for videogames and high-powered video editing. While the central processing unit (CPU) of a computer can be used for mining as well, GPUs are widely considered a better choice as they are already designed to run the same operation repeatedly for a prolonged period of time which makes them a great fit for the hashing process needed to verify new blocks.

The large increase in price that ether has seen throughout 2017 means that there are more people mining on the Ethereum blockchain than ever before. This, in turn, means you may find it difficult to find a quality GPU if you don't want to purchase used hardware which is never recommended. To counter this trend, you will simply want to frequent eBay and Amazon until new shipments come in directly from vendors. While this means you will likely need some extra time to get certain parts, it is certainly doable with a little perseverance.

The most popular cards when it comes to mining ether is going to be the NVIDIA GeForce GTX 1080 Ti which offers the greatest individual hash rate per card of any card on the market, though it also draws a comparable amount of power for doing so as well. The GeForce GTX 1070 or the AMD Radeon RX 480 are both more midrange cards that offer a more balanced mix of horsepower and power consumption. If you are willing to do some tinkering under the hood, you will actually find that the 1070 can be altered to increase its performance significantly while actually reducing its overall power consumption. When this is done correctly it will result in a card that has the optimal watt to performance ratio of any card on the market.

You can often find the GTX 1070 founder's edition online for around $450. Using MSI Afterburner you can then increase the memory interface clock to 650 MHz while also reducing its power consumption target to 66 percent. Doing so will decrease board power and heat output by the maximum amount possible without hampering performance for what you need it to do in the slightest. Doing so will also drop the temperature of the GPU by more than 15 degrees while raising the hash rate from 27.24 MH/s all the way to 31.77 MH/s. Assuming you are paying 10 cents per kilowatt hour, doing so means you can expect to make about $140 per month at the current cost of ether. This is per GPU and the amount earned is likely to continue to increase along with the price of a single ether.

When you are choosing your motherboard, it is important to pick the one with the most PCIe slots as possible as doing so will make it easier for you to run numerous GPUs at once at some time in the future. One thing it is important to keep in mind is that you are always going to want to keep an eye out for a note in the motherboard literature that says it was designed specifically for mining. These types of boards are generally going to cost more but are going to perform better, and hold up to the type of stress that mining places on a system far better than standard motherboards. They are also typically designed to add additional power for the GPU and also have more connectors to help draw as much power as possible.

Perhaps more importantly, they don't require shared PCIe lanes for external hardware like some alternatives will, which will come in handy when it comes to shoving in as many GPUs as possible. They are also going to be able to withstand the types of temperatures that poorly ventilated rigs often reach, especially if they aren't optimized. Despite all the specialization, they are also more basic than other types of motherboards which means you should expect to pay more no more than $200 for something that is extremely reliable.

Power supply: The power supply unit (PSU) is an often-overlooked component of many mining machines that nevertheless is going to require some thought to ensure that you don't find yourself without the power you need to run your machine at maximum efficiency. First and foremost, this means you are going to need to calculate the exact power requirements that your machine is going to need so you can choose a PSU that will be able to supply it all, plus a little more.

Even if you are only planning on running a single GPU to start, it is often better to purchase a larger PSU to start, rather than having to upgrade it whenever you add a new GPU. You are also going to need to know the number of power connectors that you will have available in order to ensure that you have enough space for the extra GPUs as well. Don't forget to count the number of 4-pin Molex connectors you will need as this could easily be important later.

In most scenarios you are still going to need to add in at least one splitter adapter as a means of squeezing out the maximum number of connectors possible. While you are going to want to avoid this solution if possible, as it can lead to overloaded wires if you aren't careful, if it is done correctly it can add a significant amount of additional capacity to a machine. Just be aware that you will most likely have to replace the connections multiple times throughout the life of the machine. If you choose to go down this route then you will need to take extra precautions, including checking the wires by hand on a regular basis to ensure they are not warmer than their neighbors.

If you go with the ASRock H61 ProBTC motherboard you will need to use extenders in order to ensure that everything works properly. The motherboard comes with one extra 4-pin connector but the easiest choice in the long run will be to go with powered PCIe risers instead of trying to pull the extra power through the motherboard directly. This is especially true if you plan on running several GPUs.

Its PCIe slot that is held in reserve for a video card on the motherboard can provide up to 75W of power, with much of its consumption going to the 12V line with a ratting of max 66W. While this means that having a maximum of 4 GPUs isn't going to overload your system, you are still better off not taking any undue chances, especially as the system is regularly going to be running under an extreme load.

The most popular PCIe riser is the X1 PCIe USB 3.0 and it is known for being reliable and well-designed. While these risers come with a USB 3.0 option, they don't actually take advantage of the USB interface and are instead simply an easy way to move data around in ways it needs to go. The risers require a 4-pin Molex power connector for each GPU that will be drawing from the PCIe slot on the motherboard. This means it should be well within the capabilities of any 4-pin connector that comes with a rating of 10-11A per wire at minimum.

Assuming your video card is going to max out its draw from a 12V PCIe slot line at 5.5A, about 66W, then you will need a wire that is at least 18 AWG to ensure your needs are met, as long as it is rated for at least 6 Amps. If this is the case, then you will only need to connect a single additional 4-pin connector to the line that runs to the power supply. This will not always be an option, however, as most newer power supplies automatically come with numerous 4-pin connectors that come connected to a single line. If you find yourself with this issue then you will need to ensure that no two of your PCIe extenders are powered by the same cable to ensure you remain close to the recommended specifications.

CPU: Unlike other parts of the system, the CPU that you choose is going to be largely irrelevant. The GPUs will do a large majority of all the work that the machine is going to be tasked with which means that as long as your CPU can keep your machine running, it fulfils the sum total of its job requirements. If you went with the suggested motherboard, then an Intel Celeron G1840 ($70 on Amazon) will be enough to meet your needs and more. If you decide on an AMD GPU instead, then you will want to go with a cheap Sempron CPU instead.

RAM: Much like the CPU, the amount of RAM you use is going to be of far less importance than it would be with a traditional computer as the GPU memory that you have available is what's going to matter in the end. In order to run

Windows 10, you only need four gigs of RAM but for $50 you can easily get eight gigs of DDR3 ram on a single stick so there is no reason not to have a little extra.

Hard drive: Continuing the list of things you don't need to worry much about, the only things on your hard drive are going to be the operating system and the mining program software which means a 60-gig solid state hard drive, also $50 on Amazon, will cover all your needs.

Something to keep it in: The case that you choose to keep your mining rig in is going to be extremely important as heat concerns are going to build up quickly when you add a second GPU to a standard PC case, not to mention a third and a fourth. While you can purchase a case that can support as many as six GPUs for around $200 on Amazon, you can build the same thing yourself for a fraction of the cost.

The following frame can be constructed using little more than standard milk crates, and can be easily added to, assuming you plan on expanding at a point in the future.

To create the case, you will also need a knife that can easily cut through the milk carton plastic along with a solid drill. You will also want to purchase the following:

$5 power switch

$2 PVC pipe (1-inch)

$1 metal screws (6)

$5 SurfaceGard Adhesive Bumper pads (1 pack)

$10 plastic milk crates (2)

To start assembling your case, the first thing you are going to want to do is to add the adhesive pads to the bottom of the inside of the crate and then set your motherboard on top of them. When positioning, it is important to ensure you have easy access to all of the relevant ports. This may require a bit of cutting to ensure you can get at everything you need to. With this done, you can then select the PCIe ports that are going to receive riser cables.

With this done, you will then need to place the PVC pipe into the crate so that it sits above the motherboard so that your GPUs can sit on top of it. You will want to cut the pipe so that there is roughly 2 inches sticking out from either side of the crate. On each end you are going to want to drill a pair of holes and secure it using a cable tie to prevent it from moving around unduly. This is a crucial step when it comes to the long-term stability of the system. Failing to do so could easily cause the entire system to literally fall apart.

Once the pipe has been secured, the next thing you will need to do is add in your GPUs, resting each bracket on the crate's lip when you do. You will then need to secure each GPU to the crate for the same reasons as those outlined above. You will likely find that this process is expedited if you first mark the holes you need to drill before you get ready to drill them. Next, you will mount the power switch somewhere out of the way in the crate. Finally, all you need to do is add the power supply, along with the hard drive, to the second crate, connect everything to the power supply and see if it successfully powers up.

Chapter 11: Sell Ethereum

To sell Ethereum at any of the above exchanges you must have Ethereum in your exchange wallet. If you have chose to store Ethereum on an exchange (we do not recommend this) then you will already have access to selling it. If you store your Ethereum in a wallet where you control your private key, then you will need to transfer your Ether to your exchange wallet in order to sell it. We provide further details on Ethereum transactions here however a quick overview can be found below.

At your chosen Ethereum exchange, find your Ethereum (ETH) wallet. Do not use a Bitcoin or other cryptocurrency wallet as sending Ether to these addresses could result in permanently lost funds.

Copy your Ethereum address; this starts with the characters "0x".

Go to the wallet where your Ethereum is stored and send your Ether to your exchange wallet using the copied address above.

Once your transaction has received enough confirmations (1 confirmation every ~15 seconds and the number of confirmations required varies from exchange to exchange) you can trade your ETH for a fiat currency or another cryptocurrency. Further details of selling Ethereum at an exchange can be found below.

How to sell Ethereum?

The method for selling Ethereum varies from exchange to exchange however the general process remains the same. If you are using a platform like Coinbase (recommended for beginners), liquidity is provided by the exchange and ETH sales are done directly between the seller and the platform. If you are using an exchange marketplace like Kraken or Poloniex then the sale of ETH is done between yourself and a matched peer(s).

Sale to an exchange

There are advantages to selling Ethereum directly to an exchange platform like Coinbase. Two of the key advantages are speed and simplicity; the platform will guarantee to purchase your Ethereum at a set rate. There will be a maximum amount of ETH that can be sold at any one time, however the order will be fulfilled immediately.

Adversely however, the exchange will often buy Ether at a less favorable rate, charging a small premium to cover the added risk of providing liquidity to their entire userbase. Exchanges like these will also cap the amount of Ether that can be sold in any single time period.

Sale on an marketplace

A marketplace simply connects willing buyers and sellers together. Buyers choose how much they are willing to pay for ETH (the "Bid" price) while sellers set a price at which they are willing to sell ETH (the "Ask" price).

The Ask price is always higher than the Bid and the difference between the two is known as the "spread". Sales on a marketplace have two key advantages over sales to an exchange platform: the price (both for buying and selling) is more favorable than the alternative, and the amount that can be sold at any one time is far greater. Marketplaces are generally a better choice for more experienced Ethereum traders, however the key factor is "liquidity". If a marketplace has low liquidity i.e. very little volume for a certain currency pair such as ETH/USD, then the price and speed of sale can be poor.

Marketplace sales can be difficult for certain currencies where volume/liquidity is low. For example, the cryptocurrency pair ETH/GBP is poorly catered for, often requiring users to sell ETH for USD (a highly liquid with lots of buyers and sellers) and then converting the received USD to GBP at the bank following a SEPA/SWIFT withdrawal. To find the most liquid exchange for your chosen currency pair, see the Ethereum markets at CoinMarketCap.

Limit/Market order

Selling Ethereum on a marketplace is more complex than selling directly to a platform. When selling ETH the seller has two options for how they wish to sell. Either the seller creates the market (market maker) and specifies the price at which they are willing to sell, or they sell to an existing buyer who has listed a buy (Bid) price. Market makers create a "limit order" which specifies the lowest price that the seller is willing to sell

Ethereum for. These markets are then matched with willing buyers. Alternatively – and more simply – a seller can sell using a "market order", stating that they are willing to sell to the highest buyer on the market (market taker). However, with a market order there is a danger that the seller may get an unfavorable price as explained below. Sellers who act as a market maker are often rewarded with lower trading fees than market takers.

Example: the ETH/USD market has 2 buyers, one who is willing to buy 5 ETH at $790 and another who is willing to buy 10 ETH at $780. If a seller chooses to sell 8 ETH (using a market order), they will receive 5 ETH at $790 and the remaining 3 ETH at $780. If however, they choose a limit order at the same price, then they would sell 5 ETH at $790 and the remaining 3 ETH would not be sold until a willing buyer joined the market.

By creating a limit order, the seller guarantees the price at which their Ether will be sold.

Placing a market order in the example above may be acceptable for many sellers. However, consider an example where the seller wished to sell 10,000 ETH. The market would be liquidated at a lower and lower price, potentially selling some ETH tokens for just a few dollars. This has been seen before – resulting in a "flash crash". Generally speaking, a seller will be better off selling using a limit order, however the mechanics of this should be understood properly before placing a sell order. One disadvantage of a limit order is that the sale may take hours or days to be executed in full. If the price moves negatively, a limit order may need to be closed and reopened at the lower price. A market order will guarantee a fast sale on a highly liquid exchange and is often preferable for those willing to sacrifice profit in return for speed.

How to short Ethereum

"Shorting the market" is the act of selling an asset today and buying it back at a later date – for a lower price (subsequently making profit). An Ethereum trader who chooses to short ETH/USD would benefit from the price of ETH/USD falling. For example, they would sell ETH/USD today for $700 and buy back later for $600. In this case, the possible upside of a short position is limited by the price of ETH/USD falling to $0. Note that the short seller does not actually own Ether, instead the cryptoasset is borrowed and thus creates a liability.

The possible downside to short selling is unlimited – the price of ETH/USD may rise indefinitely creating limitless losses. For this reason, shorting Ethereum is very high risk, and positions are typically opened and closed over the short term with stop-loss limits put in place (automatic closure of a position if losses exceed a user-specified amount).

Whilst very few people have dared to short Ethereum/Fiat currency pairs, some may choose to short ETH/BTC, expecting that the price of Bitcoin will outperform the price of Ethereum over a given period. Conversely, a trader may choose to short BTC/ETH – predicting that the opposite will occur.

Shorting a cryptocurrency can lead to losses that exceed the user's deposit. For this reason, to short Ethereum the trader would require a margin account. A margin account ensures that there are additional funds (a buffer) in the trader's account which will cover losses if the position moves adversely. If the margin account is not maintained (that is to say, if there is not enough of a buffer), then the position may be closed automatically by that platform. This is known as a "margin call".

Ethereum can be shorted through 3 different avenues.

Regulated futures markets

CME Group and CBOE have begun trading Bitcoin futures and allow investors to short BTC/USD and other fiat currency pairs. It is anticipated that these same futures markets will be opened up to

Ethereum. At this point in time, selling or "shorting" Ethereum on these markets is not yet accessible.

Cryptocurrency exchanges

There are several cryptocurrency exchanges which allow users to open margin accounts. These exchanges will allow users to sell ETH/fiat and ETH/crypto pairs as well as many other non-Ethereum pairs.

Contracts For Difference (CFD)

One of the easiest ways to short Ethereum today is through a CFD broker. These brokers allow users to essentially "bet" on the direction that the price of Ethereum will move. If the trader believes Ethereum will fall, then creating a "sell" order will reward the user if the price drops. An advantage to using CFDs is in the highly regulated nature of the broker (see the top Ethereum brokers above) and the simplicity of setting up an account makes this option highly appealing to new traders.

Why invest in Ethereum?

There are several reasons why a user might choose to buy or invest in Ethereum, here a handful of examples.

Buying Ethereum as an investment

Accessing token sales and other blockchain investments

Hedging against the incumbent fiat system

Diversifying a traditional portfolio

Buying Ethereum for use

Interacting with blockchain-based IoT devices

Using smart contracts and the EVM

Paying wages internationally

What investment strategy?

Investment strategies vary, and suitability is subject to your own personal risk tolerance. This guide is for information purposes only, and if in any doubt consult a financial adviser.

Buy and hold

One of the most common investment strategies for Ethereum is "buy and hold". If Ethereum is to replace even a fraction of fiat currency, its value will be far greater than it is today. The same can be said if Ethereum becomes the currency of choice for the "machine payable web" which will enable billions of devices to transact value efficiently with each other.

Given the volatility of Ethereum, those looking to buy may want to consider "dollar cost averaging"; spending the total investment amount in chunks over X period of time to acquire Ether at an averaged price.

Buy and diversify

It is safe to say that predicting the future of Ethereum is much like predicting the weather in 5 years time. It is unlikely that Ethereum will disappear anytime soon, but as Ethereum has shown Bitcoin, it is possible for a little-known cryptoasset to become a dominant force in a short period of time. Purchasing Ethereum to exchange for other cryptoassets like Ripple (XRP) and Ethereum Classic (ETC) is a good way to hedge against the unexpected failure of any given coin. Whilst one coin may fail, many VCs and technologists are in agreement that cryptoassets of some nature will become ubiquitous in the future.

Ethereum Wallets

If you're looking to buy Ethereum and would like to secure your funds yourself (recommended), then understanding how Ethereum wallets work is critical.

Light client Ethereum wallets:

MyEtherWallet

Jaxx.io (also supports other cryptoassets)

Private key

Now that you have chosen a particular Ethereum wallet, it is important to understand the private key that will be generated with it before depositing any funds. The private key is the key to your wallet; if anyone else has your private key then they also have full access to your wallet and its associated funds. When creating a wallet, you will be asked to take a copy of your private key. The above wallets generate your private key offline, it is never sent to a server and therefore cannot be intercepted. It is now up to you how to store and backup your private key. Many users choose secure cloud storage with 2-factor authentication or offline prints of their private key. For larger sums of Ethereum, extra security measures can be taken with "hardware wallets" described further below.

Transactions and addresses

You have now downloaded an Ethereum wallet and secured your private key. Before funding your wallet with Ether, it's important to become familiar with the make-up of a simple Ethereum transaction.

Your Ethereum wallet will automatically generate a handful of receiving addresses (also known as "public keys") which are a function of your private key. Unlike the private key, receiving addresses can be distributed freely without risk of theft, and payments to these receiving addresses will add funds to your private key (wallet). For those interested in the technicals, see Bitcoin's "public key cryptography".

All transactions on the Ethereum blockchain are publicly visible through a "block explorer" such as Etherscan.io. As an example, we are going to look at this arbitrary transaction of 0.2 ETH. In this transaction you can see 2 public keys:

From:

0xea674fdde714fd979de3edf0f56aa9716b898ec8

To:

0x3d167984ae0868194ffd97759ff74e342ff3140d

MyEtherWallet example transaction

Using MyEtherWallet as our example software, the above transaction is simply the input of the address we wish to send funds to, the amount, and the gas limit (fee). The latter is set automatically by the software – but it's prudent to double check the cost as miscalculations have been known to occur.

The from address does not need to be specified. This address is chosen automatically based on the balance of each address.

Once a transaction has been sent, a transaction hash is created and shown to you in the software. This transaction hash can then be put into a block explorer and the same details we have just looked at can be found for your new transaction.

Block height and confirmations

This is the mined block which your transaction was included in. It takes roughly 15 seconds to mine a block. The period between your transaction being broadcast (i.e. sent to a receiving address) to when it is first included in a block is a period in which your transaction will be "pending". Inclusion in a block is called a confirmation, and every subsequently mined block adds another confirmation. The more confirmations a transaction has, the more "bedded-into" the blockchain it is. Transactions with 30+ confirmations are generally considered extremely secure – they will persist in the blockchain for all eternity.

Block height can also be thought of as the block number since the creation of the blockchain. In our example's case, the transaction was included in the 3,804,203rd block to be mined.

Receiving transactions

Now that you are familiar with sending Ether, the same idea can be applied to receiving it. In the case of purchasing Ethereum, once it has been bought from an exchange, the exchange's withdrawal function will ask for your wallet address. Input your address along with the amount of Ether you wish to withdraw to your wallet, and then once confirmed, a transaction hash will be shown. The Ether will immediately show as being "pending" in your wallet, and you can follow the number of confirmations it has using the transaction hash on Etherscan.io.

Transacting Ethereum safely

Whilst rare, there have been several horror stories of users losing thousands of dollars in Ether from poor due diligence. Here are a few of the key items to check off when making a transaction of a significant sum.

Copy and paste the address

Never type in a wallet address by hand. Addresses are long and case-sensitive, a single mistake will result in the funds being lost forever. There is no charge back or customer support number in Ethereum.

Check the transaction fee

A good Ethereum wallet will show you the calculated transaction fee in dollars and cents. Always double check that the transaction fee is reasonable.

Check, double and triple check the address
Once you've copied and pasted the address which you wish to send or receive Ether to, check it over and over until you're certain it's correct. Looking at the first and last several letters/numbers will ensure it's been pasted correctly.

Good wallet software will also confirm the address that you are sending or receiving to. This mitigates the risk of malware intercepting and replacing the address you input.

Test your transaction
One of the driving forces behind Ethereum adoption is the low transaction fees. There is no harm in sending a negligible amount of Ether in order to test your understanding of the process and that all of the details are correct. This will ensure that everything goes smoothly when sending larger amounts.

Securing Ethereum – the easy way
This section of the guide intends to simplify the process of securing Ethereum

for non-technical users. This method of security hands over management to the exchange with which you purchased Ether. In this instance, users can secure their newly purchased Ethereum by leaving it in their wallet on the exchange itself. This introduces some risks, including platform risk (the platform may fail as has been seen before with the MtGox Bitcoin exchange) as well as the risk of digital theft, as was seen with Bitfinex. Ultimately, some users choose to secure large sums of Ether by leaving it on an exchange, but in doing so they give up their ability to audit and ensure its security. The decision to do this is a personal one – how much do you trust the exchange, and how much are you willing to leave in the hands of that exchange? Whilst it is unlikely that your coins will be stolen, or that platform will become insolvent, it is a real possibility that should be accounted for.

A note on exchanges

Securing Ether on an exchange is a legitimate approach to security, particularly if the funds stored are relatively small in size compared to an investor's overall portfolio. However, when storing coins on an exchange, you do not own the private key. Essentially, you have handed over responsibility of your Ether to the exchange. Exchanges are not the same as a bank, and the same financial regulations do not apply. Insolvency or theft may result in lost funds.

Securing Ethereum – the hard way

Securing Ether is a critical step in ensuring that your investment is safe. Unlike many developed countries, the banks will not protect your cryptoassets like they protect your cash. As mentioned, investing in cryptocurrencies in unforgiving, securing Ether properly is critical. Those looking for a simpler security option by handing over this management to a 3rd party can see the above section.

Securing Ethereum through a hardware wallet

Hardware wallets are one of the safest ways to secure your Ether. Hardware wallets generate and store your private key offline, and at no point is the private key exposed to your connected device (PC). Storing your coins offline in this way mitigates the risk of digital theft – one of the most common attack vectors for cryptoasset holders. As with other Ethereum wallets, a recovery seed is provided on creation, and a PIN is chosen to secure access to the device itself. It is the PIN and the recovery seed that must then be secured extremely well, as access to either by a malicious individual may result in loss of funds. Further protection can also be taken in the form of 2-factor authentication and multi-signature wallets as discussed below.

The two most respected hardware wallets for Ethereum (and other cryptoassets) are Trezor and the Ledger Nano S. Those storing Ethereum on a Trezor device will need to use it in combination with MyEtherWallet (see the full guide here). For that reason, many users opt for the ease of use that comes with the Ledger Nano S.

Additional security measures you can take
2 Factor Authentication

Whether you choose to store your coins on an exchange, desktop/mobile wallet or hardware wallet, 2 Factor Authentication (2FA) is a highly recommended additional security layer. The 2FA process requires that the user inputs a one time password (OTP) before being able to login to a wallet or send Ether. Google Authenticator is one of the most popular interfaces for 2FA and is used by a range of Ethereum wallets.

Different wallets and exchanges will implement 2FA in different ways, however the additional security that it provides remains the same. A potential thief would not only require your password to steal your Ether, but access to the physical device from which the OTP is generated as well.

A word of caution

2FA through an app like Google Authenticator has so far proven extremely secure. However, some platforms choose to bypass the use of an app and instead send an OTP over SMS. SMS 2FA should be avoided entirely, as the OTP can – in many cases – be observed without needing to unlock a phone. More catastrophically, social engineering has been used to convince telecoms staff to port a phone number to a new SIM; if an attacker is able to do this, then the phone number alone can be used to gain access to any platform "protected" by SMS 2FA.

Multi Signature Wallets

These wallets allow the user to secure their Ethereum by requiring multiple participants to sign each transaction. Typically a multi signature wallet is "2 of 3", meaning that 2 of a total of 3 private keys must sign the transaction for it to be successfully broadcast to the network. In these instances, the 3 private keys can be split across different physical locations along with their own physical security to ensure that there is no single point of attack.

Chapter 12:How To Buy Ether

In this chapter, you are going to learn various methods through which you can buy Ether, and how to go about using each of these methods. Keep in mind that before you get started on the process of buying Ether, you first need to get yourself an Ethereum wallet, you can go with any of the options discussed in Chapter Five.

The simplest and most straightforward method of buying Ether is to go through a cryptocurrency exchange. These are online platforms that facilitate the trading of cryptocurrencies between users. Owing to Ethereum's popularity and huge market cap, there are several cryptocurrency exchanges that support Ether, which means you have multiple options regardless of your location. While different cryptocurrency exchanges have slightly different procedures, the process is of buying Ether through a cryptocurrency exchange involves the following:

The first thing you need to do is to create an account with your exchange of choice. This step will involve providing some information about yourself, such as your name and email address. Before signing up for an account, it is important to confirm that the exchange has support for your country and the fiat currency you intend to use to purchase Ether.

Once you have signed up for an account, most exchanges will require you to provide some additional information before you can make deposits or withdrawals. In most cases, they will ask for your government issued identification, a photo of yourself and a proof of address. This is done to ensure that the exchange is in compliance with Anti-Money Laundering (AML) and Know Your Customer (KYC) laws.

Now that you have verified your identity and address, you can go ahead and choose your preferred deposit method. Different exchanges accept different deposit methods, so you need to check this before signing up for the account. This

135

information can easily be found on the exchange's website, along with the fees charged for each method. Some of the most common deposit methods supported include wire transfers, PayPal payments, SEPA and credit/debit cards.

Once you identify the deposit method that works best for you, you can now go ahead and deposit your fiat currency into the exchange platform. Dollars and Euros are supported on almost every exchange, while many other exchanges support other major fiat currencies such as Sterling Pounds, Canadian Dollars, Japanese Yen and Chinese Yuan. Your deposit might take anywhere between a couple of hours to a few days to reflect in your exchange platform account, depending on your chosen deposit method.

Once your funds reflect in your exchange platform account, you are ready to buy Ether. The process varies depending on your chosen exchange, though most exchanges try to keep the process simple and intuitive, even for beginners. After receiving the Ether in your exchange platform account, it is always advisable to transfer them to a wallet whose keys you control. Do not leave them on the exchange platform.

Popular Cryptocurrency Exchanges Where You Can Buy Ether

Coinbase

This is the world's most popular cryptocurrency exchange, and a good option for you if you want to buy some Ether from the USA, Canada, the UK, Europe and Singapore. In addition to Ether, Coinbase supports several other cryptocurrencies. Coinbase has been in operation for over six years and has established itself as one of the most reliable and trustworthy exchanges. Buying Ether through Coinbase is easy and straightforward. If you are a more experienced user, you can also use Coinbase's GDAX, which offers a more advanced set of features. Coinbase allows users to deposit funds (fiat currency) through bank transfers, SEPA and credit/debit cards. However, if you opt to use credit/debit card payments, you will have to contend with a much lower limit. Despite being the most popular cryptocurrency exchange, Coinbase is only available in 32 countries. Therefore, you need too ascertain that your country is supported before signing up. It's also good to note that Coinbase does not sell Ethereum Classic.

CEX.io

This is another popular exchange that that has been around for a while. The platform is registered with Fincen and provides brokerage services in addition to being a crypto exchange. Cex.io started supporting Ethereum in 2016 after they brought their cloud mining service to an end. Unlike Coinbase, Cex.io is available worldwide. You can deposit your funds on Cex.io through wire transfers, SEPA or credit/debit card. Getting verified gives allows you to buy a higher amount of Ether through credit card. The Cex.io website is intuitive and user-friendly, which makes the buying process quite easy, even for beginners. On the flip side, the fees charged by Cex.io are a bit on the higher side. However, it is still a good option for those who live in countries not supported by Coinbase.

Coinmama

Founded in 2013, Coinmama is a cryptocurrency exchange and brokerage service that supports Bitcoin and Ethereum. One of the greatest advantages of Coinmama is that you can buy up to $125 worth of Ether without being verified. This means you can go to their website, sign up, and complete your purchase in less than 20 minutes. Coinmama only allows you to buy Ethereum using credit or debit card. Their customer support is quite excellent, and the service is available worldwide. However, their fees are a bit high.

BitPanda

Formerly known as Coinimal, BitPanda is an Austrian cryptocurrency broker that allows people within the Eurozone to buy and sell Ether. Founded in 2014, the platform has gained a lot of popularity among users in Europe. BitPanda allows you to pay for Ether through SEPA, credit card, Skrill and several other payment methods that are popular in Europe. Their fees are fairly low, and their website is quite fast and secure

Gemini

Founded in 2015 by the Winklevoss twins, the New York based cryptocurrency exchange has rapidly grown in popularity. Gemini offers its services to users in North America, Europe and Asia. The Gemini platform works like a traditional forex exchange, allowing users to trade with each other directly, with prices being determined by the users. Gemini only supports deposits made through bank transfer. One of its greatest advantages is that its fees are low.

Changelly

Founded in 2016, Changelly is still a fairly new entrant into the market. Despite having not been around for long, it has also become quite popular. One of the best features of Changelly is that it allows users to trade one cryptocurrency for another. This makes the platform a good option if you already own some Bitcoin that you want to exchange for Ether. While it is possible to buy Ethereum from Changelly using fiat currency, the fees are quite high. If you have some Bitcoin that you want to exchange for Ether, the process will take you about 30 minutes.

Bitfinex

Bitfinex is a Bitcoin exchange that also allows users to trade cryptocurrency pairs. This means that, if you own some Bitcoin, you can exchange it for Ether on the platform. Bitfinex does not support fiat currency, therefore the only way of acquiring Ether through the platform is to buy some Bitcoin first. Bitfinex is quite secure, with advanced security mechanisms to keep customer information safe. Users' funds are also stored in cold wallets to eliminate the threat of online hacking attacks.

Kraken

This is another US based cryptocurrency exchange that allows you to buy Ether using fiat currency or in exchange for other cryptocurrencies. Kraken is one of the very first exchanges to be established, having been founded in 2011. Kraken supports more cryptocurrency pairs than Gemini and Coinbase. If you decide to buy Ether from Kraken using fiat currency, you can make deposits through bank transfer. The only complaint about Kraken is that their user interface is poor. Their customer support is also said to be a bit slow.

Purchasing Ethereum Anonymously

Like I mentioned earlier, purchasing Ether through an online exchange will require you to provide proof of your identity. However, sometimes you might want to purchase Ether anonymously for one reason or the other. In this case, you can do so by purchasing Ether from peer to peer exchange platforms, such as localethereum.com.

Localethereum.com is a platform that brings together Ethereum buyers and sellers within the same geographical region. You can think of it as an online marketplace, just like eBay, only that the product in this case is Ether. The buyer and seller can agree on a payment method that works best for both of them whether that is bank transfer, PayPal, Skrill, credit card, Bitcoin, or even cash. For its efforts, localethereum.com charges the seller a small percentage of the trade. This is the most effective method of buying Ether anonymously. However, you might need to take some precautions. For instance, if you decide to meet up with the seller to exchange Ether for cash, you need to think about your own personal safety.

Alternatively, you can purchase Bitcoin and then exchange your Bitcoin for ether on shapeshift.io. This is an exchange that allows you to exchange different cryptocurrencies without having to create an account with the platform. However, you cannot buy a large amount of Ether from shapeshift.io.

Finally, you can purchase Ether anonymously from an Ethereum ATM. These are ATM machines that allow you to exchange cash for Ether. Ethereum ATM's do not require any form of identification. All you need to do is to select the amount of Ether you want, enter your Ethereum wallet address and insert money into the machine. The Ether will be automatically sent to your wallet. However, since Ethereum ATM's do not ask for proof of identity, you can only buy small amounts of Ether from an Ethereum ATM.

How To Choose The Best Ethereum Exchange

With so many available options, choosing the best Ethereum exchange for your needs can be a bit hectic. To ensure that you are making the right decision, below are some factors you should consider before settling on a specific exchange platform.

Supported location: Some exchanges like Coinbase and Gemini only offer their services in specific geographical regions. Therefore, before registering for an account with an exchange platform, you need to ensure that their services are available in your geographical location.

Deposit methods: You also need to confirm whether an exchange supports you preferred deposit method. For instance, if you want to make your deposit via PayPal, you cannot use Gemini. You should also keep in mind that different deposit methods attract different fees. Generally, the faster and more convenient the deposit method, the higher the amount of fees you can expect to pay.

Supported currency: Different exchanges accept deposits in different currencies. In most cases, this depends on location. Most exchanges accept USD and Euro deposits. However, if you intend to use any other currency, it is always good to first check whether the currency is accepted on your preferred exchange. For instance, if you intend to deposit Japanese Yen, you can only use an exchange that accepts Japanese Yen deposits. Similarly, you cannot use Bitfinex if you intend to buy Ether using fiat currency.

Security: Buying Ether involves money therefore you need to be sure that you can entrust your funds and personal/financial information without the risk that it might get stolen by hackers. Check the exchange's website to find out the kind of security measures they have in place to protect your funds and information.

Support: Sometimes, you might face some challenges setting up your account or buying Ether, especially if you are a beginner. You want an exchange that has a responsive and helpful support who will quickly help you solve any issues that might arise.

Fees: Different cryptocurrency exchanges have different fee structures. You should look for one that offers you the lowest fees.

Reputation: Finally, before settling on a specific exchange, it is always a good idea to check its reputation. What are other people saying about it? Check different online forums and review websites to see the kind of reputation the exchange platform has. If you come across several customer complaints, this should act as a red flag.

The above are some of the considerations you should have in mind when choosing a cryptocurrency exchange. However, sometimes it is impossible to find one that meets all your requirements. In such instances, you might have to sacrifice on some of these factors. For instance, if the only exchange supported in your country charges high fees, you might have no other option but to go with it or forego buying Ether.

Chapter Summary

In this chapter, you have learned:

The simplest and most straightforward method of buying Ether is to go through a cryptocurrency exchange.

Most cryptocurrency exchange platforms will ask for your government issued identification, a photo of yourself and a proof of address in order to remain in compliance with Anti-Money Laundering (AML) and Know Your Customer (KYC) laws.

Different exchanges accept different deposit methods, so you need to check this before signing up for an account with a cryptocurrency exchange.

After receiving Ether in your exchange platform account, it is always advisable to transfer them to a wallet whose keys you control. Do not leave them on the exchange platform.

Some popular cryptocurrency exchanges where you can buy Ether include Coinbase, CEX.io, Coinmama, BitPanda, Changelly, Gemini, Bitfinex and Kraken.

If you want to purchase Ethereum anonymously, you can do so by purchasing Ether from peer to peer exchange platforms, such as localethereum.com. You can also purchase Bitcoin and then exchange your Bitcoin for Ether on shapeshift.io.

You can also purchase Ether anonymously from an Ethereum ATM.

To choose the best Ethereum exchange, you should consider factors such as supported location, supported deposit methods, supported currencies, security, fees, customer support and reputation.

In the next chapter, you will learn about Ethereum wallets.

Chapter 13: How To Make Money From Ethereum

Now that you have a great understanding of how to mine Ethereum, it's time to learn how to make money from it. The value of this particular cryptocurrency has been at an all-time high, and it keeps rising every time. Just to put it into perspective, its value almost doubled in February 2017 and there is a massive demand for it with more people getting into the market.

Trading versus Buying Ethereum

Whether to invest or not to invest in Ethereum is a personal choice that depends on whether you are risk tolerant or risk-averse. As with every investment, it has its risks because the future value is not 100% certain. Digital currencies may be very risky, but the greater the risk, the greater the return.

Those who invested in Bitcoin during its early days are now reaping the rewards. The best thing about Ethereum is that it costs less than Bitcoin and would thus be easier to invest for people who do not have so much money to spend.

Here are the reasons why you should consider trading Ethereum:

Appreciating prices: Though there are predictions that the value will keep going up, no one has any idea on up to what point. Therefore, the trick is to buy now and then cash out before it crashes or collapses.

Stability: Ethereum has been growing organically over the years and there have been very few spikes so far. Although it is not exactly predictable, this is a great sign of its potential.

The idea behind it: The developers of Ethereum want users to see it as a giant computer software where all kinds of applications can run. This is precisely what has ensured that many companies all over the world are backing and supporting it. This fact means that it will be steady for a while.

Its popularity: After Bitcoin, it is the most popular digital currency in terms of size and hype.

Here are some of the ways that you can make money from Ethereum:

Mining Ethers: You can choose to mine ethers, which are the units that help you access the Ethereum system. Mining is for tech-savvy people who are very familiar with the systems that we talked about in chapter three. You will have to pay some initial costs but you will make money in the long run if the trend stays the same. However, to make a considerable amount when mining ethers, you must ensure that your computer has the capacity to handle it. While some people prefer to mine for Ethereum on their own, it is usually better to form a pool of miners and work as a team. This way, you will have a much higher return.

Trading Ethers: After acquiring some Ether, you can then trade them. To make a profit, you need to buy at a low price and then sell at a higher price. Since the price tends to appreciate, you can buy them, wait for a while, and then resell them. You can buy Ethers at Coinbase through your credit card, but of course, there are some verification processes to ensure security when completing your bank details. Coinbase also has a great referral and reward system. It shows you the price of Ethers so that you can be sure before you purchase. For example, the Coinbase platform informs you of the current Ether price and then predicts the projected price for the end of that year. However, this is just a prediction and no one can know for sure whether it will hold up.

Freelance Blockchain Programmer: Although this is something far from the scope of this book, if you decide to research and apply yourself, you can learn how to become a blockchain programmer. Programmers currently earn a lot of money, and this situation will definitely improve as demand continues to rise.

Writer/ Blogger/ Consultant: When you achieve success in Ethereum, you can offer your expertise to people and share what you know about it. You can also become a consultant if this is something that you love to do.

Of the above options, look at the one that makes the most sense for you. During the initial stages, mining and trading can be the best ways for you to make money. The rest of the options are just things that you can do later on as you progress.

With all the information you have learned in this chapter of the book, you are now ready to mine or set up your wallet to trade in Ethereum.

Chapter 14: Ethereum Key Players

And Technical Infrastructure

The technical infrastructure of the Ethereum project mirrors the center Ethos of the job itself by being widely decentralized. The
Ethereum project is a huge undertaking, led primarily by its developers, but also depends on the dispersed efforts of its diverse
community. The key players in this industry are:
· Developer Leads
· The Ethereum Foundation
· Decentralized Projects
· Ethereum startups
Let us investigate these players.
Developer LeadsMembers of the Ethereum community generally Incline towards being evasive when talking its early history and growth
despite the fact that it's an open secret that there have been a number of changes to its membership. As an example, the first

thread which introduced the Ethereum project on the Bitcoin Talk online forum has been altered since its original publication,

with the full list containing the titles of the developers and architects that have since changed onto other endeavors.

A major distinction with Bitcoin is that while The Bitcoin creator--Satoshi Nakamoto--abandoned the project at an infancy stage,

Ethereum has definitely been fueled by the active participation of its inventor, Vitalik Buterin, and also the core development

group. Two of those core developers within the Developer Leads which are often cited are Gavin Wood (previously the project's lead

C++ developer) and Jeffrey Wilcke (also the lead Golang developer). Other notable developers include the numerous people employed

from the Ethereum Foundation.

The Ethereum FoundationThe Major company behind the Ethereum

Undertaking Is the Ethereum Foundation. Launched as a non-profit
making organization in June 2014 at Switzerland, the Ethereum Foundation has made concerted efforts to help drive its development
by integrating innovative approaches towards development.

Its membership includes of the following:
· Vitalik Buterin, the founder of Ethereum.
· Gavin Wood, formerly the job's lead C++ developer.
· Jeffrey Wilcke, its direct Golang programmer.
· Bernd Lapp
· Stefano Bertolo
· Yessin Schiegg
· William Mougayar

Decentralized ProjectsThe Ethereum project has already seen the Emergence of several projects that seek to bring its fundamental
concepts to life. One of these projects are:
· The DAO
· The Augur

StartupsThough the Ethereum project remains in Its first phases, the initial wave of Ethereum projects has been observed with keen
interest from venture capital firms that have experience in the Blockchain domainname. One of the Ethereum startups are:
· Backfeed
· BlockApps
· Ether.camp
· Ethcore
· Otonomous
· Akasha
· Colony
· ConsenSys
· Plex.ai
· Provenance
· Slock.it

Chapter 15: Ethereum Versus Bitcoin

At this point in the book, we have discussed all of the major aspects of Ethereum use from a computational and userability perspective.Now that you know about all of the major elements of Ethereum and how to access all of the major aspects of it, we will now turn our attention to focusing on the current climate of blockchain technology from a business perspective.This chapter will focus on how Ethereum differs from Bitcoin, and will also discuss what the current climate is surrounding Ethereum versus Bitcoin.

Bitcoin Giving Blockchain Technology a Name

When bitcoin was first introduced, it was seen as an innovative way for people to make payments and send money online without the use of a middleman or any third party interference.Overtime, however, there have been known problems with bitcoin technology.These problems include hackers infiltrating the bitcoin system and stealing thousands of dollars in bitcoin from its users.In the past, bitcoin has been known to attract people who are looking to sell products on the "black market".This means that criminals have looked to bitcoin as a way to sell products that are illegal in nature.For this reason, multiple countries including Saudi Arabia, Ecuador, and Bolivia have banned the use of bitcoin within its borders.

While bitcoin has its fair share of potential problems, the fact that it was the first blockchain cryptocurrency should not be ignored or overlooked; however, similar to the VHF video tape or the tape player, once the first of a new technology has been introduced, people will likely try to better it or advance it in some fashion.This seems to be the case for Bitcoin.Today, many people recognize the limiting factors of Bitcoin.Instead of trying to better the Bitcoin application itself, many innovators have instead sought to better the technology behind Bitcoin.This is the blockchain.It's important to understand that the problems that exist within bitcoin are problems that are intrinsic to the Bitcoin platform running on top of the blockchain.

Bitcoin as a Springboard for Ethereum Development

In addition to being the first blockchain cryptocurrency to exist, Bitcoin also can only function when currency is being traded between various parties.This is where Ethereum primarily differs from Bitcoin.As we already know, Ethereum expands the functionality of cryptocurrency by allowing people to not just trade currency with one another, but also trade goods through Smart Contracts or subcurrencies that the individual Ethereum user can create on his or her own.It can be argued that this makes Ethereum and Bitcoin similar in one important way – each application is the first of its kind underneath of the cryptocurrency umbrella, respectively.Bitcoin was the first currency trading blockchain platform, while Ethereum was the first beyond-currency trading blockchain platform; however, this comparison should not exclude the fact that without the development of Bitcoin, Ethereum would likely not exist.This fact positions Bitcoin as an integral aspect of blockchain technology development, regardless of Bitcoin's found shortcomings.

How Bitcoin and Ethereum Differ

Even though the importance of Bitcoin as a starting point for Ethereum should not be overlooked, there are still key differences between the two applications.Let's take a look at some of these differences now:

Difference 1: It's Intended Use

When Bitcoin was first developed, it's main purpose was to serve as a coin that could not be regulated by the federal government.If you remember, it was developed during a time when the regulatory banks could not be trusted as a money-governing entity.Ethereum's purpose is to serve as a "world computer".To this end, it's intended purpose is multifaceted and can prove to be useful to many different types of people with many different end goals.

Difference 2: Its User Base

Because of its intended use, it's been proven that both Ethereum and Bitcoin cater to different types of audiences. Users of Bitcoin are typically looking for a network that is completely devoid of authority. There is no central "administrator". On the other hand, Ethereum requires that its users pay a central administrative entity through a small mining transaction fee in order to use its platform. While Ethereum is charging more for users to use its services than Bitcoin, it's important to remember the types of people who have been known to gravitate towards bitcoin. When there is an administrative entity that can dictate what's going on within the blockchain, it's likely that less riffraff or foulplay will occur.

Difference 3: Transaction Pace

The time it takes for a single Bitcoin blockchain to be uploaded to the Bitcoin network is ten minutes. The time for a single Ethereum blockchain to upload is ten seconds.This has been a problem for Blockchain in the past, because this makes it more difficult for the application as a whole to process an infinate number of transactions efficiently.

Difference 4: The Code Type

Bitcoin is designed for the computer programming language known as C++.Contrastingly, Ethereum uses a language known as Turing-Complete.This language is comprised of a whopping seven differently programming language types, including Go, Rust, Haskell, Javascript, Java, Python, and C++.This difference speaks to the universality of Ethereum, because programmers with all different types of backgrounds are able to operate within it.

Difference 5: Evolving Hash Functions

As we already discussed, both Bitcoin and Ethereum currently use POWs to protect their networks; however, Ethereum is currently working towards operating on what's known as a Proof-of-Stake (PoS) blockchain.The reason why Ethereum is seeking to move towards a PoS blockchain is because the cost associated with running a POW blockchain is rather high and inefficient.Additionally, for the lower price that it costs to run a PoS, the developers of Ethereum believe that they will be able to provide their users with a higher level of security than a POW can provide.

Difference 6: Ability to Grow

From its onset, Bitcoin was not developed to grow at the fast rate at which it did.This has led to various problems within the Bitcoin application, most notably, it not being able to keep up with the demand for the product.On the other hand, Ethereum was designed with growth in mind.Currently, Bitcoin is trying to play catchup in this regard.

The most important difference between Ethereum and Bitcoin is arguably that you can implement Smart Contracts using Ethereum, but that has already been discussed in this book.Hopefully, this chapter has provided you with valuable information regarding not just how

Ethereum and Bitcoin differ, but also the current climate of how Ethereum plans to evolve and how Bitcoin seems to be falling a tad behind in a few key ways.

Chapter 16: Financial History Of Ether

Let us study the price history of ether over the years. In 2014, ether was introduced in the market. Of course, as can be expected of any other new altcoin, it did not have any significant value. It started out slow. In 2015, ether reached its $1 milestone and started to draw more attention from a few cryptocurrency investors and traders. It was unstable up to the end of 2015 with its price going and fluctuating below a dollar. However, in the latter part of January 2016, it started to cross the $1 milestone again and got even more established. In April 2016, it already crossed more than $10. It continued to grow slowly. Of course, it was also subject to the usual fluctuations in price that is inherent in the cryptocurrency market, but you could easily recognize that its price was slowly but constantly increasing. At the beginning of 2017, it experienced a loss that its price dropped down to $8.

However, starting that period, the price of ether started to grow quickly. In April of the same year, its price reached $48. Within three months, its price rocketed to $283. Before the end of 2017, its price was close to $1,000. As of January 14, 2018, the price of 1 ether is around $1,430. Since it continues to draw more attention and interest in the market, many professional investors claim that its price will continue to grow. All known cryptocurrency exchanges now include ether in the list of its cryptocurrencies being traded. In fact, some experts agree that 2018 is most likely the year that ether will take the place of bitcoin as the number one cryptocurrency in the market.

This is the financial history of ether up to January 14, 2018. The good news is that when you read the current news on the cryptocurrency market, there are so many positive news pieces about ether and Ethereum in general. There are also many new altcoins that are based on the Ethereum platform. Needless to say, this also boosts the price of ether in the market. Although the future of ether remains unknown, there are good reasons to believe that this 2018 will most likely be the biggest hit and price leap that Ethereum will make. Considering relevant factors, including the current position of Ethereum in the market and how it is perceived as a cryptocurrency, there is more than 75% probability that this 2018 is the best moment to invest in Ethereum.

Of course, just like investing in any other cryptocurrency, it is important for you to do your research and have a closer look at the market. You should never underestimate the high volatility of the market. Therefore, even though circumstances show just how highly profitable investing in ether would most likely be this 2018, make sure that you do your own research and analysis of the market, and make your investment wisely.

Just like other cryptocurrencies, ether is also subject to the usual fluctuations in price in the market. This is normal and is considered as innate part of the cryptocurrency market. However, the important thing is to see its price behavior in a long-term perspective. It cannot be denied that Ethereum has been growing steadily and significantly these past weeks to the point that its growth appears to be unstoppable. Another good news is that the more that it grows, the more attention it draws to itself, and so the more investors are willing to make an investment in Ethereum. This, of course, will push its price even higher than it already is. Unlike other assets, cryptocurrencies like Ethereum work like social media in the sense that the more it establishes itself and broadens its network, the stronger and more profitable it becomes. Indeed, there is so much to expect from Ethereum, especially this 2018. Although this is still a mystery to be uncovered, many believe that Ethereum is going to dominate the market this 2018.

Chapter 17: Future Of Ethereum

Ethereum is continually developing and changing, and if you are not sure about investing with Ethereum, then you need to look at all of the new developments that are coming from the system and how much it has to offer you when it comes to how you can make money with it.

It is alright to be hesitant about investing with a new platform that is continually evolving because you do not know what it will do and, of course, you do not want to lose money to something that is not going to be around.

However, think about when the stock market first came around. People were hesitant about that too, but, it is still around to this day, and people are continually making money or losing it with the stock market.

It is all a game of figuring out where your niche is and how you can improve your skills so that you can beat the odds and make a profit. Just remember that you need to learn from your mistakes!

And, just because you do not make money right away does not mean that you should give up with Ethereum. Keep investing because you are bound to make money sooner or later! Especially with Ethereum developing the way that it is!

Vyacheslav Putilovsky, an analyst at Expert RA, a Moscow rating company, it is a highly developed banking market, and the leading banks want to match if not overtake their western competitors in adopting such technology. Blockchain technology can be used to verify intellectual property rights, contractual agreements and public ledgers without intermediaries.

The consortium of U.S. banks in May raised $107 million from its members, which include HSBC Holdings PLC and Bank of America Corp.

Prototypes Available Decentraland has prototypes available for public testing. It offers unlimited possibilities for virtual content creators, including games, art, education and health care applications.

Developers can create and monetize applications on the platform. They can use MANA to buy LAND and other goods and services on the platform. LAND's value is based on its proximity to high-traffic hubs, as well as its ability to host in-demand applications. A property owner can host other users on their LAND and determine how they interact with the virtual world around them. While gaming, art, and social applications are anticipated within Decentraland, the protocol enables users to come up with other novel use cases.

Secondary Market Trades Decentraland has teamed with district0x, a collective of decentralized communities, to provide users the ability to trade LAND in the secondary market. While unclaimed LAND can be bought at the same exchange rate (1,000 MANA = 1 LAND), differences in traffic and positions could allow LAND parcels to trade at different prices on a secondary market

Chapter 18: Mining Ether On Ethereum

You do not have to buy ether if you do not want to. You can always mine it and here is how.

Put C++ visual on your computer

You will either be working with a 64-bit or a 32-bit computer. You will have to download the proper visual from Microsoft when you are downloading C++ to mine appropriately. If at all possible you will want to work with a 64-bit computer because the 32 system has some issues when you are mining ether. But, if that is all that you have to work with, you will have to be patient with the program as it works through its issues.

Install ethereum

The next step will be to download Mist. Mist will be your graphical interface that will be where your wallet is for ethereum. Mist is user-friendly and offers help through a browser application that you will want to get so that you can find other people who will be in a position to help you should you need help.

Get blockchain

This step will take a while for you to complete due to the fact that it will be where you download the blockchain onto your computer. The file that you will be downloading will be a ten-gigabyte file. So, if you start downloading this record, it is a good idea to find something else to do while it does what it needs to do.

Set up your wallet

Now will be a good time for you to set up your wallet. You have a lot of options when it comes to picking which wallet you can use. You will need to look at the security that the wallet has to offer as well as the benefits that you will receive from that wallet to choose the one that will be the one for you to use. Have a wallet is vital because it will be where all of your ether will be sent when you are rewarded for the work that you do.

Install Nvidia Cuda, CL SDK, or AMD

You will need to check your GPU before you can install the program that will be working off of the GPU. The application that you download will be based on your GPU's strength.

Install AlethOne Miner

You are most likely working as an individual so you will want to download AletheOne.AletheOne is the system that you are working on to allow you to mine. Remember that a 32-bit system will freeze and present other issues when it comes to mining so try and avoid them.

Wait for the DAG initializationWhen AletheOne isdone, you will have to wait for around ten minutes so that you can build a DAG. The DAG file will be stored on the computers RAM so that an algorithm can be created that is ASIC resistant.

Join a mining pool

The chances are that you are not going to be working in a warehouse that is full of GPU, so you will want to find a mining pool that will be right for you. This will go back to the competition that you read about earlier. You do not want to join a pool that is not going to make it to where you have to fight against too many miners. This will be because if there are other people that are more competent than you, you will be putting yourself in the position where you may not be getting rewards for the work that you do or, if you do get a reward, it is not going to be as big as it could be.

Chapter 19: How To Buy, Sell, And Store Ethereum

The thought of using new technologies tend to intimidate most of us. The same goes with using Ethereum. But once you get over the fear, it can be rewarding.

In its current form, Ethereum is not intuitive but anyone owning a computer or even a smartphone should be able to try the platform if they have 'ether'. These are unique bits of code which allow updating of the ledger's blockchain.

Ethereum Wallets

Before you even deal with ether, you need a secure place to store it. This is called your Ethereum wallet which also stores the private keys. Be aware, though, that losing the private key poses a bigger problem than forgetting a password. Ether is forever lost when you lose the private key.

One of the advantages of digital currencies is the removal of third-party mediators which can slow down transactions. The problem is, when they are removed, there's no one to turn to when you need help in recovering your private key.

There are quite a few options for storing cryptocurrency – mobile wallets, desktop wallets, paper wallets, and hardware wallets. Each one has advantages and disadvantages when it comes to security and convenience. More convenience means less security and vice versa.

Mobile Wallets

Also referred to as light clients, mobile clients can connect to the network and do transactions with less downloaded data, making them suitable for downloading on a smartphone.

Although the mobile client is quite convenient, it's not as safe as full Ethereum clients which validate transactions autonomously and don't rely on nodes or miners in sending them the correct information.

Private keys that are stored on devices not connected to the Internet tend to be more difficult to hack. This practice is best employed when storing large amounts of ether and is also referred to as cold storage. This method, however, is not as convenient to use on devices connected to the Internet like a desktop computer or a smartphone.

Desktop Wallets

You need a PC or a laptop computer to run desktop wallets. An Ethereum client which is the entire copy of the Ethereum blockchain needs to be downloaded first. Different programming languages can be used to develop Ethereum and performance depends on the language used and how the code was streamlined.

The download will usually take a couple of days but it will depend on your Internet connection speed. The download time will grow as Ethereum grows. You also need to sync the content of the wallet with that of the latest blockchain transactions.

Paper Wallets

Another option for cold storage is to simply print the private keys on a piece of paper, thus the name paper wallet. You can also carefully handwrite the private key if you want. You then need to store it somewhere safe like a vault or a deposit box. Online tools can be used to generate key pairs on a computer and not on the website's servers. This could potentially make keys vulnerable if someone hacks the site.

Hardware Wallets

This option is considered to offer both security and convenience in perfect balance. Hardware wallets are typically the size of a finger and can be used to sign transactions without getting online.

Buying Your First Ether

Buying ether differs by currency or by country. Basically, you buy it from someone who has ether through online or personal transactions. The option to buy ether in person is currently rare with only New York and Toronto is known for frequent meetups related to Ethereum trade. It's less likely in areas with a small population.

Most of the time, you need either US dollars or Bitcoin to directly buy ether. Using other currencies might involve extra steps. Bitcoin, being the most popular digital currency in use nowadays, is preferred by most people who want to trade their ether. So if you're buying ether using rubles, you first need to buy Bitcoin using the currency and then trade it for ether. Once you get your ether, it can be sent to another person directly through peer-to-peer. A small transaction fee might be required and this amount goes to the miners.

You can usually find the best prices for cryptocurrencies on trading platforms as compared to going through brokers. Expect a bit of delay before you can finally buy ether, however. It's not unusual for account verifications to take a few days. Here are some trading platforms to get your ether from:

-Coinbase – It isarguably one of the world's most well-known places to trade Ethereum. You may find that prices are higher compared to other trading platforms. They lead the market when it comes to buying experience so they're the most preferred by those just beginning to dive into the cryptocurrency frenzy.

Coinbase best works if your plan is to buy and hold ether and not trade day-to-day. They've undergone some downtime during hectic trading recently, but they are currently building a more efficient system.

-Kraken – This is another globally known trading platform and has been more successful in dealing with sudden influxes of cryptocurrency interests compared to less known alternatives like Bitfinex and Poloniex. Kraken is widely popular in the UK and in Europe.

-BTCMarkets – The only real option for Australian digital currency traders. Ether prices are quite high but there are no other practical ways to get it from down under. The good news is that you can sell it at much higher prices.

What to Do With Ether

Now that you have ether, what's next? Bitcoin is similar to Ethereum as far the previous discussions in this chapter are concerned. But Ethereum applications have disadvantages over that of its big brother.

Once you have ether, you can create or join smart contracts. These are codes that are executed automatically when the terms of the agreement are met avoiding the use of an intermediary. Smart contracts can also be bundled to develop decentralized applications. People who have ether can join into or use these applications.

How Does it Work?

Like the rest of cryptocurrencies, Ethereum has a confusing buying, storing, and selling system. Let's compare it to a system that we are all familiar with.

The string of numbers embossed or printed on the front of a credit card is used by banks to identify where to send funds when it is swiped or used to buy online. With digital currencies, you can generate a similar identification number that indicates where the funds can be debited from.

The identification number consists of two main components – a public key, and a private key. These keys are strings of letters and numbers and are linked together using cryptography.

You send the public key to the other party for them to know where the money should be sent. If you want to receive ether from other people, you will need a set of scrambled letters and numbers which has been derived from the same public key. This will be your address.

When you need to spend your ether, you use your private key to sign over funds and this key is similar to your password. Using the credit card analogy, this is the same as your pin code that you type in before you can withdraw money from an ATM.

Using digital currencies might be more complicated than using credit cards but the benefit is autonomy. Banks need to approve the bank account application first before giving you the credit card. Using open blockchains, like Ethereum and Bitcoin, you can generate your own identification number any time you want.

Chapter 20: Ethereum - Decentralized Application (Dapp)

The life of most developers revolves around learning new languages, platforms, and even frameworks. More interesting is when a developer learns a completely new paradigm. One of the newest and technologically different paradigms is the blockchain decentralized network.

Because it is a completely different paradigm, we are going to look at some technologies that are needed in the consensus network, and also look at what makes the creation of a network.

Main technologies

●Hash Function - Cryptographic

Hash functions take a piece of information and map it to a piece of data of a specific size. For example, a 2MB file that is passed via a hash function produces two hashes of 128 bits size in length. A Cryptographic Hash Function performs the functionality and fulfills 3 major requirements:

➢No information is provided, depending on the non-reversible hash that was produced by the input data.

➢Minor changes in the change of input produce an output hash that is different, in a way that the hash can only be calculated using the hash function.

➢An extremely low chance that 2 different inputs can produce the same hash.

Public Key Cryptography

This is an encryption class method that needs two different key creations; the "private key" is only for the owner and the "public key" is used by anyone. There are several attributes that are useful when it comes to key cryptography.

1. The encryption of data by anyone who uses a public key and uses a private key to decrypt it.

2. The private key holder's ability to sign using a private key and the information verified by the user who has a public key, without the uncovering of the information. This is normally used in a DCN's account's

systems. It is used to form the fundamentals of transmitting transactions.

P2P Networking

In this network, computers are connected directly to one another without requests being sent to servers. The computers in the network are referred to as peers, and they all have the same standing as the other. This network relies on altruistic nature of the peers, and they share all the resources that are available on the network.

Technologies in Crypto Economy

The Blockchain

It is similar to a database type that is used in a DCN. Information can be held, and rules are set depending on the information update. Essentially, it is updated in blocks that are chained together by the use of hashes that originate from the block content of the previous block. A block has the current and historical information. Requests or Transactions are used to change the database state, and the blockchain stores the signature.

Proof of work

Proof of work was previously a prevention mechanism. In essence, it is a simple way of proving that a large operation has been done. Many times, it is implemented by hash functions (cryptographic). If you are provided with some data part, (e.g., block header and list of transactions), you have to locate the second data part, which after combining it with the first, a hash that has some characteristics (e.g. some zeros that are trailing) is produced.

Technologies in Ethereum

EVM (Ethereum Virtual Machine)

This is an important innovation designed to be used by everyone in a p2p network. The EVM can read/write both executable data and code, participate in the verification of digital signatures and execute code in a manner that is quasi-Turing. When a message is verified using a digital signature, and when the blockchain information assures that it is okay to do so, then the code will be executed.

The Generalized Blockchain and Decentralized Consensus Network

As we said earlier, Ethereum is a p2p network where each peer stores a copy of the database and runs an EVM which maintains and changes the status of the blockchain database. The integration of Proof of Work in the blockchain technology makes new block creation to require the members of the network to take part in the proof of work. Incentivization of the network delivers the consensus for peers to accept the longest blockchain by the distribution of 'ether,' which is a cryptographic token.

After this, we are left with a technology that is not compatible with the p2p or client-server network, because its state is not consistent. Because of the cryptographic security, and its distributed nature, it can be a third party that is not trusted, and it does not have any interference with the outside parties. Cryptocurrency decisions have made the financial impact on organizations, people, and other kinds of software.

This has made developers come up with alternative ways of enabling how different software and components on the internet relate. We are going to use cases and explain the benefits of each decentralized apps.

Development environment setup

Web designers have an easy time designing on Ethereum because the language used in the development is familiar to anyone who knows JavaScript. There are three software parts that each developer downloads

AlethZero

It is a GUI client that has advanced features which includes force mining, private chains, and a full WebKit.

Mist

This is the DApp browser and the mining client used by the client to access DApps.

Mix

It is an integrated development environment that builds and compiles contracts alongside their frontends.

Requirement of the Software

There are three software parts which we have discussed above that need to be downloaded.

The first part is to download is the current AlethZero stable binary, a C++ client, and any operating system. If problems are experienced in the build, then stick to a more current version, which may have fixed the issues.

The second part requires you to install MIX, the integrated development environment for both Mac and Windows. You can also install in Linux.

The final part requires you to install Mist to DApps and tweak the front-ends while you are developing.

Extras

A Mix or a text editor can create the contract code at the backend that we shall be writing. When it comes to serpent, it is best if you set up your editor to save serpent contracts using the suffix '.se' as python, and save '.Sol,' for solidity. Do not use a live refresh when you are working on the front end of the HTML because their testing is not complete.

AlethZero Setup

There is a heavy development that is going on in IDE MIX, and even though there are several features in place, we shall focus mostly on front-end using the client AlethZero development which has a JavaScript console, a compiler, and peeking tools that check into the state of the blockchain.

Our samples are going to be executed on AlethZero's single chain without a connection to the network. Contract development on the test net needs to be reserved for contracts that are going to be shared by others. When alethzero is running in this mode, others may join the chain as long as they are using the same name, and connect directly using the 'connect-to-peer' mode.

The first DApp

We are going to help you build your first contract example, even though a lot of details will be assumed to get to the finished product easily. Let us look at the decentralized web applications.

Basics

The decentralized web which is referred to as 'web3' is enabled by Ethereum. The difference between 'web2' and 'web3' is that Ethereum does not use web servers, and no middleman exists to squander commissions or steal information and there is no DDoS.

A decentralized app has a front end which is written using HTML and the backend which can be referred to as the database.

If you are into the use of bootstrap, you can use the framework because DApp's frontend can access the network fully and CDN's are also accessible. DApp's frontend HTML writing is just like website development and converting to web 3 from web 2 is often insignificant.

If you are a ruby, meteor or angular fan, you will love the reactive programming that is packed in it by the use of callback functions. Another great benefit is that each DApp recognizes the pseudonymous nature of every user, thanks to Ethereum's cryptographic principles to function. In simple words, as a user, you don't have to create an account to get to the DApps, think of it as a default setting.

Installing the client

The most stable version is the master build because the client is not that stable. We are going to use the AlethZero and the C++ implementation for developers. You need to install the master which has the updated features.

You can download the Windows and OSX binaries, and follow specific instructions for Ubuntu that are available on the internet.

AlethZero Overview

After starting AlethZero, you should be able to do something like this:

The interface normally varies dependent on the resolution of your screen.

Now, at the center of the screen, there is the browser which is what we call the WebKit. Browsing can be done from the

WebKit; it's just like the normal browser. The other panes have technical information and debugging information. It is useful for the developer and another user. This look is different from 'Mist' which is the Ethereum browser. Once Ethereum is launched, it is going to have a completely different look as shown below.

You can reorganize the screen if you wish. Panels can be drag and dropped on each other to form a stack.

Choices

You can build financial applications, games, social networks and even gambling apps on Ethereum because it is a programming platform.

We are going to write a basic contract that functions as a bank, but instead, it has a ledger that is transparent enough to be audited by the whole world. 10,000 tokens will be used, and since it would not be fun to have all tokens to ourselves, a method will have to be created to send out to friends.

This is a simple way of issuing our money. In web2, it would not be possible to have such an app in MySQL and PHP, with users trusting you with the accounts.

Contract

The contract is the backend which uses Solidity language. Other contract languages exist that can be used to build Ethereum's backend, Serpent, and LLL. Solidify will be used because it is the formal language that is supported by ETHDEV.

We are going to do two things to build our small bank.

1.) Create instances of an account that has some tokens for us to start.

2.) to move tokens around, we will build a send function

Let us get to it then.

```
contract metaCoin {
mapping (address => uint balances
function metaCoin() {
balances[msg.sender] = 1000
}
function sendCoin(address receiver, uint
amount)
Returns (bool sufficient) {
if (balances[msg.sender] < amount) return
false;
balances[msg.sender]-= amount;
balances[receiver] += amount;
return true
}
```

Don't worry if you can't understand the code above; It's not as complicated as it looks. Contracts are split into methods. The 1st one is called metaCoin. It's a special constructor that stipulates the initial state of a data storage contract. Constructor functions are named just as the contracts. This code of initialization is run once after the creation of the contract. The contract code follows suit; it is the part that lives eternally in the Ethereum network. In our instance, it is a function that counter-checks that the sender has a balance large enough which if it is confirmed, a transfer of token is done to another account.

In detail,

mapping address => uint) balances;

This code creates a storage mapping where the code can write information to the storage of the contract. In this code, the mapping is defined for important value pairs that are the type address and uint defined as balances. This is the repository of the coin balances. The two data types we have addressed include uint and address.

```
function metaCoin() {
balances[msg.sender] = 10000;
}
```

This is the contract initialization which will be run once, and it does several things. One, it looks up public addresses using msg.sender, for the sender of the transaction, which is you in this case. Secondly, it accesses the contracts storage using mapping balances.

Let us look at the 'sendCoin' function that will be executed when the contract is called. It is the only called executable function. There are two arguments in the function, receiver, and amount. The receiver is a 160-bit public address and amount is tokens to be sent to the receiver.

The balance is going to be checked on the first line. If it is less than the tokens being sent, the other code is not executed. If the balance is more than enough, there will be a false conditional evaluation and the amount will be subtracted by the two lines from the balance: balances[msg.sender]-= amount;,and the balance of the recipient's account is added.Balances[receiver] += amount;.

Now you have a function that sends tokens between accounts.

There are more details that you need to learn about DApp that can be found in more advanced books. You still have to learn about how to contact storage, JavaScript API 1, 2 & 3, contracts which send transactions, variables, contracts interaction, Event Logging, Gas & Gas Price and how to use Mix.

Conclusion

Thank you for making it through to the end of this book, let's hope it was informative and able to provide you with all of the tools you need to achieve your goals whatever they may be.

The next step now is to make a determination on how you are going to invest in Ethereum and other cryptocurrencies. While you may think that it is already too late, this is not necessarily the case. The value of most major cryptocurrencies is expected to go up in the next coming months and years.

By the end of 2017, the prices may increase by over 10%. Within a year or two, the prices could double and things can only get better. With governments around the world approving laws and regulations to accommodate cryptocurrencies, it can only mean that these currencies will soon be traded and used to pay for goods and services across the globe.

Experts in the technology sector have said that the blockchain, the decentralized technology used by most cryptocurrencies, is the greatest invention after the Internet. These experts know exactly what they are talking about. Any wise person would wish to put their money in such a venture.

However, even as you plan on investing in Ethereum, you need to take precaution and invest wisely. The slow but sure investor wins the race. You need to take it slow, buy reasonably and keep increasing your share of Ethereum or other cryptocurrency on a regular basis. Also ensure that you keep your cryptocurrencies as safe and as secure as possible. Since you have invested your money in them, then you should take steps to keep them secure.

If you follow the advice given here and invest prudently, then there is no doubt that a couple of months or years down the line, you will be glad that you invested in Ethereum and your investment will be worth a good amount of money.

CPSIA information can be obtained
at www.ICGtesting.com
Printed in the USA
BVHW040956250621
610376BV00007B/1483